AF266844

This Book Belongs To:

____

____

My favorite element is:

____

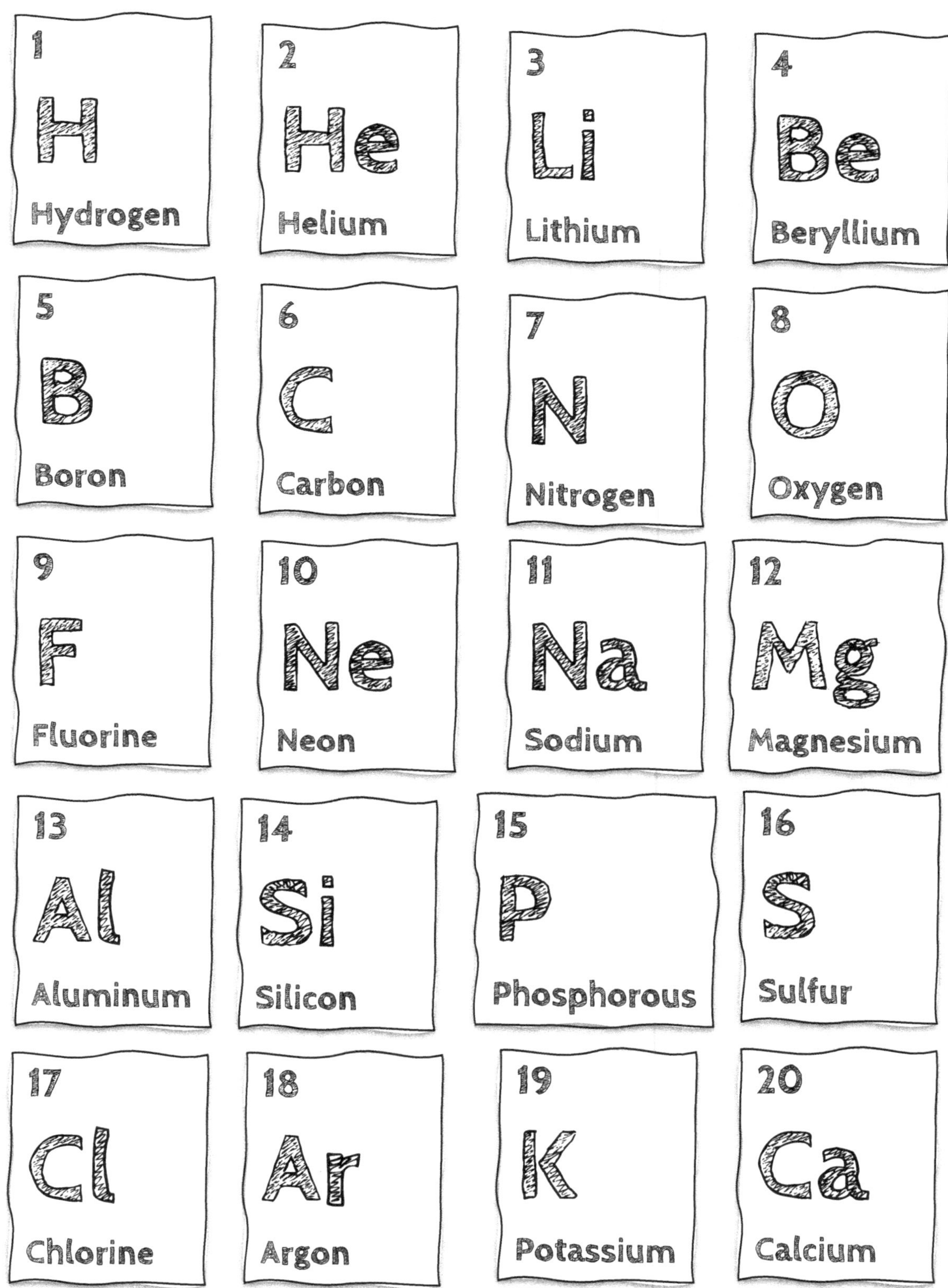

(Periodic Table of Elements – American Chemical Society, n.d.)

"Logic will get you from A to B. Imagination will take you everywhere."

— Albert Einstein

THERE ARE ELEMENTS ALL AROUND US!

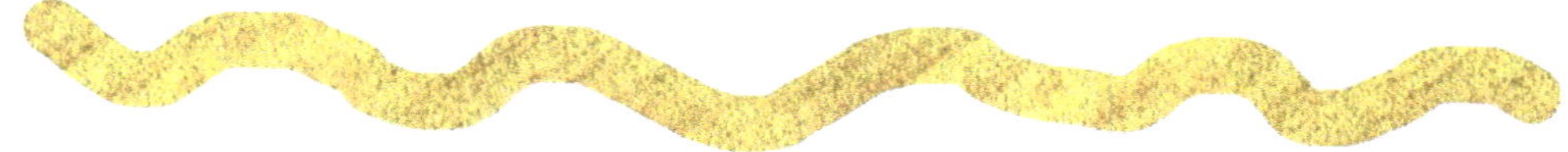

You see elements in different forms each day. They are in your home, school, food, drinks, and even toys! Use the drawing prompts to spot the elements in items that you see around you every day.

In this elemental sketchbook, the first 20 elements in the periodic table are presented as drawing prompts. Draw a picture of something containing the featured element. You can write some sentences about your drawing, or even make up a short story based on your drawing. It's up to you!

Learn about the world around you and have fun!

First, what in the world is the "periodic table"?

THE PERIODIC TABLE

Scientists make discoveries every day! A long time ago, scientists in different parts of the world discovered unique elements. Many of them worked to produce a version of a table of elements called the periodic table, which all scientists could use. Kids like you can make discoveries and be scientists too!

In 1869, a scientist called Dmitri Mendeleev was the first to publish his version of the periodic table.

Currently, scientists use the periodic table of chemical elements to rapidly look up information concerning any discovered element, such as atomic number, mass, chemical symbol, and other properties.

There are 118 elements that we know of... what a feat!

Thanks, Dmitri Mendeleev!

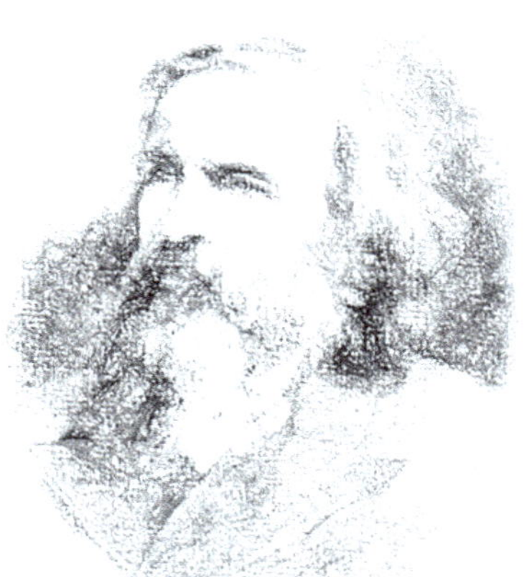

Dmitri Mendeleev
Chemist and Inventor
(https://commons.wikimedia.org/)

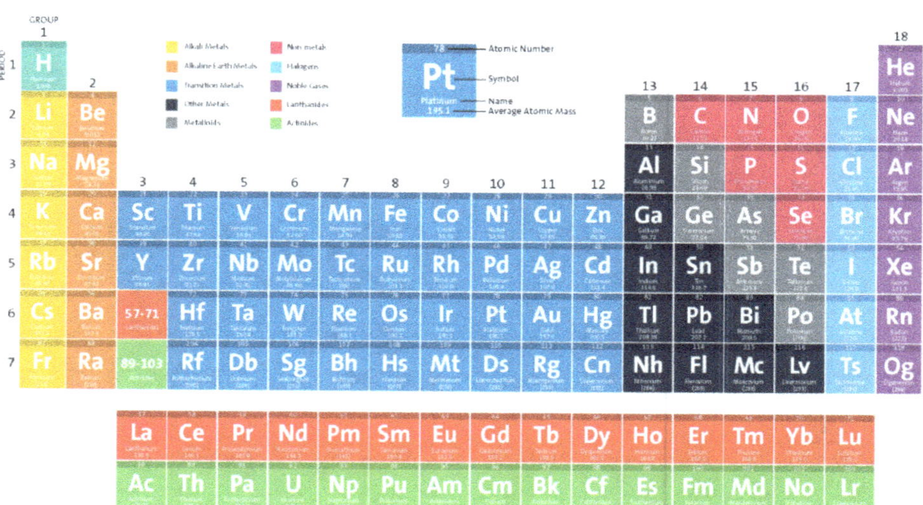

(Periodic Table of Elements – American Chemical Society, n.d.)

Use a magnifying glass to see all the elements in the periodic table.

20 Elements, tackled by one talented artist/writer...

That's YOU!

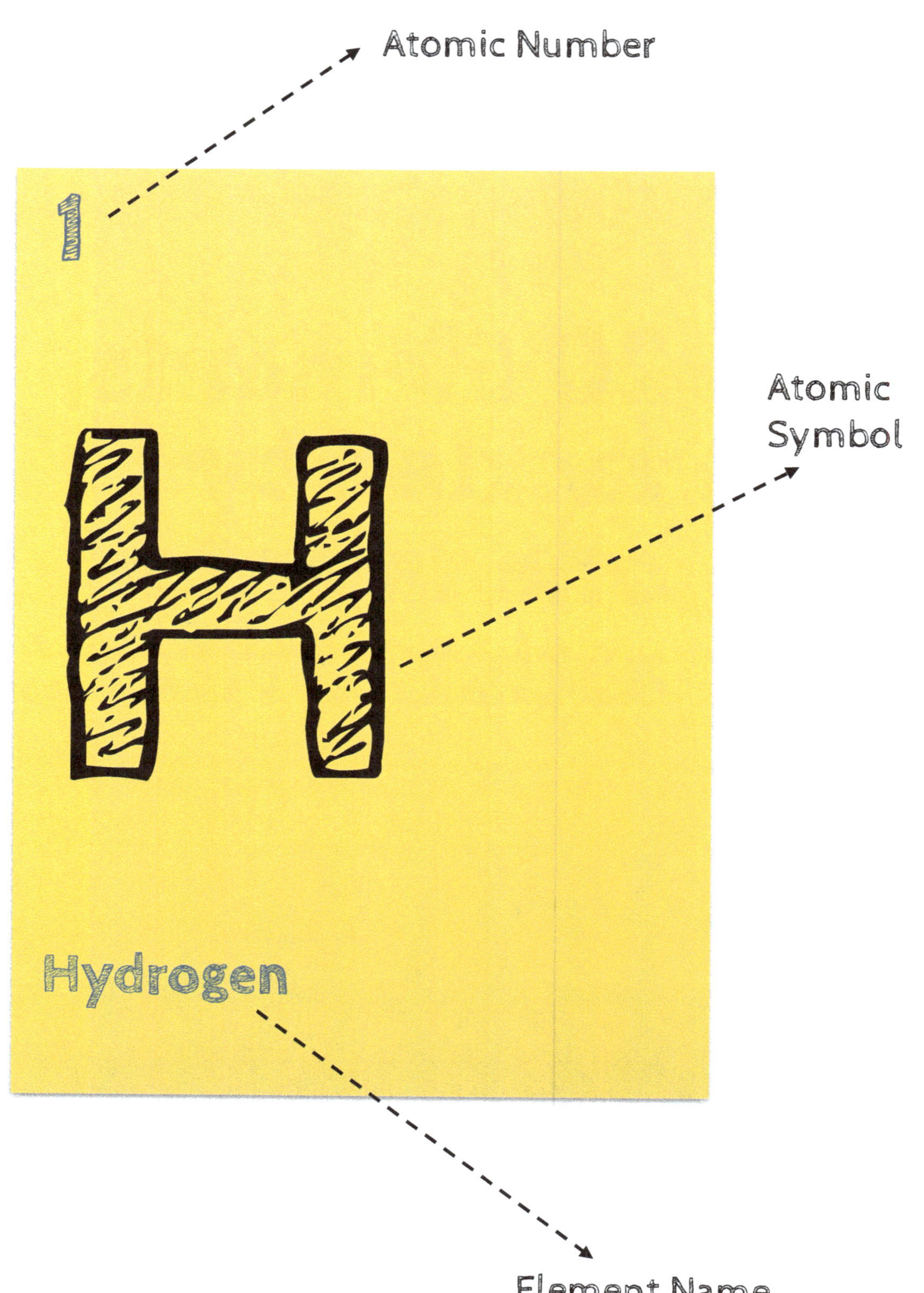

Atomic Number
Atomic Symbol
H
Hydrogen
Element Name

FACTS ABOUT HYDROGEN

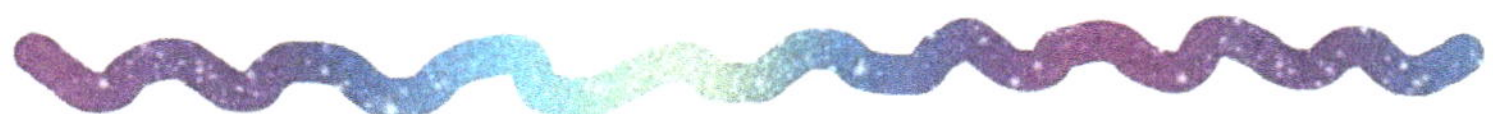

✳Hydrogen is the most common chemical element in the human body, and the universe

✳Hydrogen combines with oxygen (another element) to make water

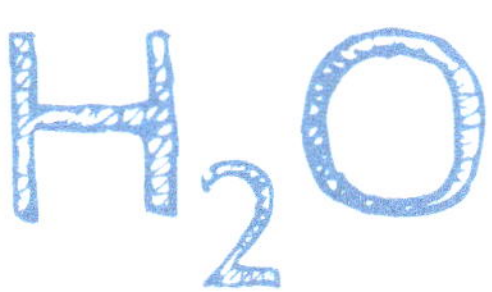

✳The Sun, Jupiter, and the stars that you see at night are mostly made up of hydrogen

(Emsley, 2001; Stwertka, 2002)

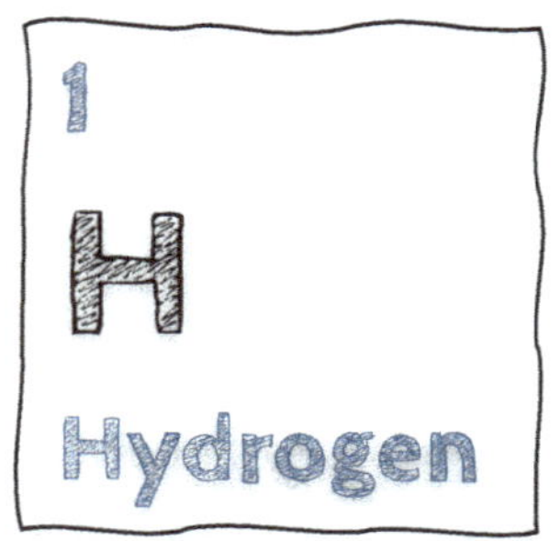

What item around you contains hydrogen?

Draw it!

Write a short story that features the
hydrogen element.

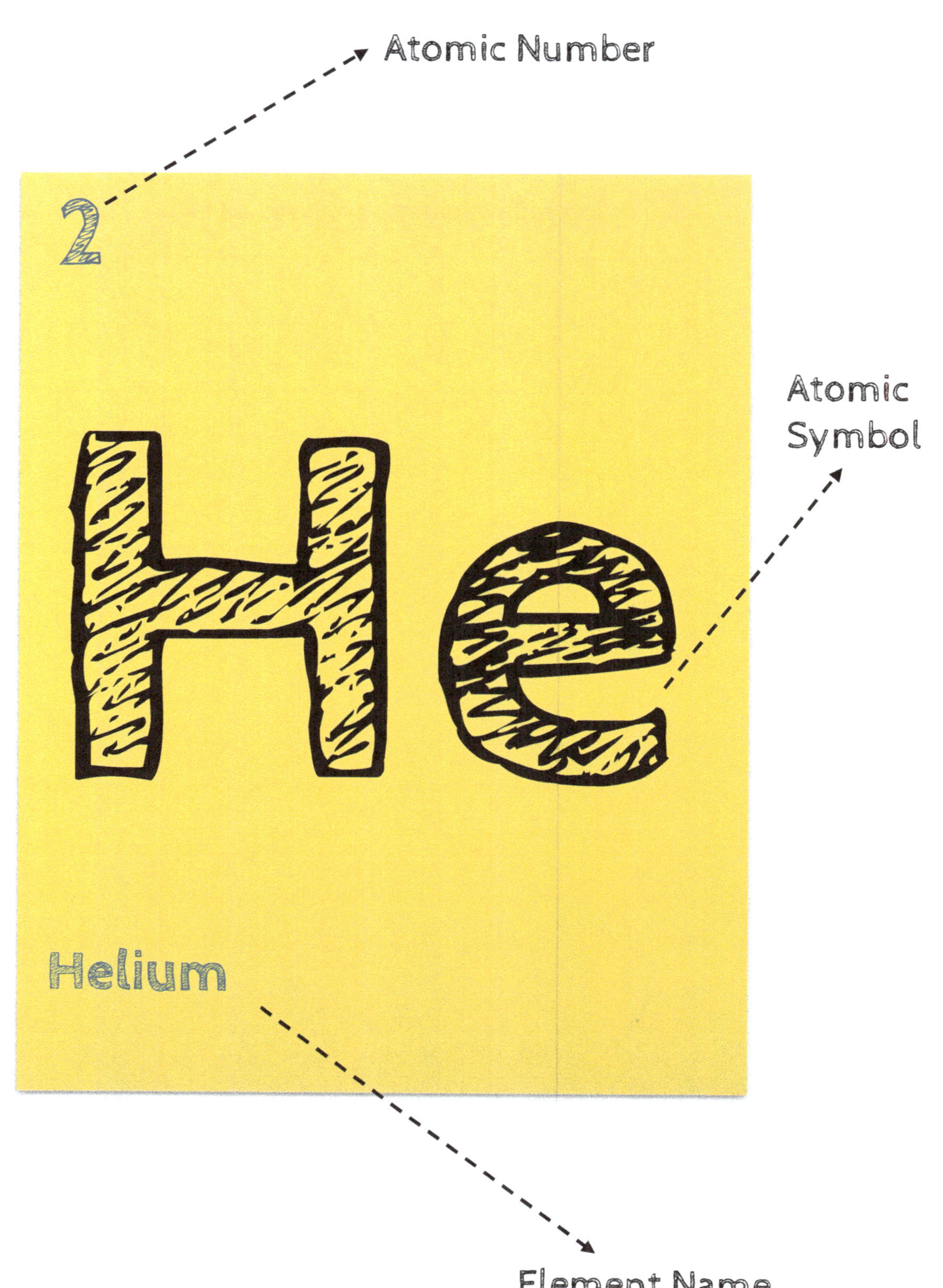

Atomic Number
2
He
Atomic Symbol
Helium
Element Name

FACTS ABOUT HELIUM

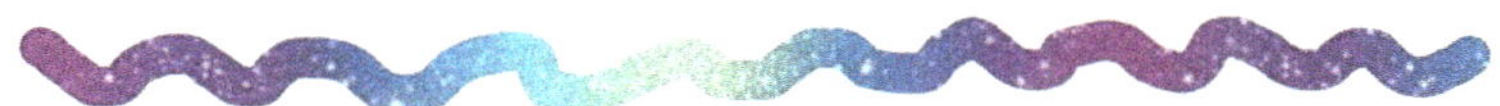

* Helium is the gas used to fill party balloons

* The name Helium originates from the Greek word "Helios" (Sun)

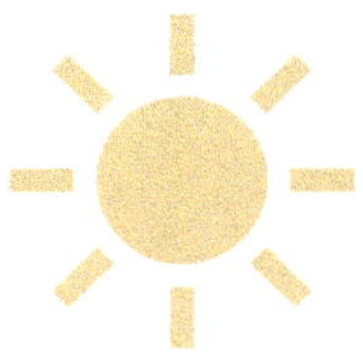

* Helium is a noble gas, which means that it is not normally chemically reactive

(Emsley, 2001; Stwertka, 2002)

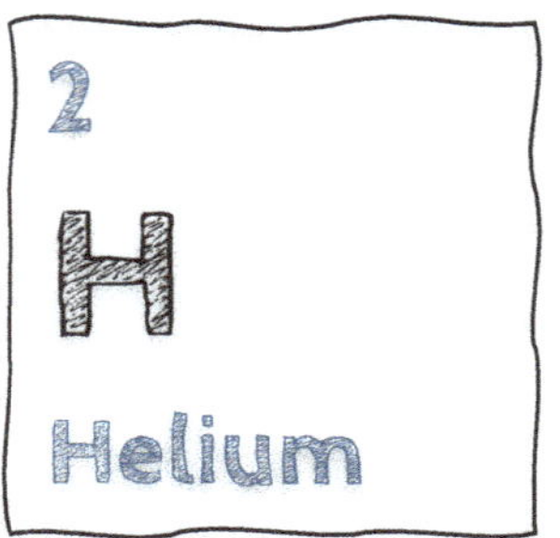

What item around you contains helium?

Draw it!

Write a short story that features the
helium element.

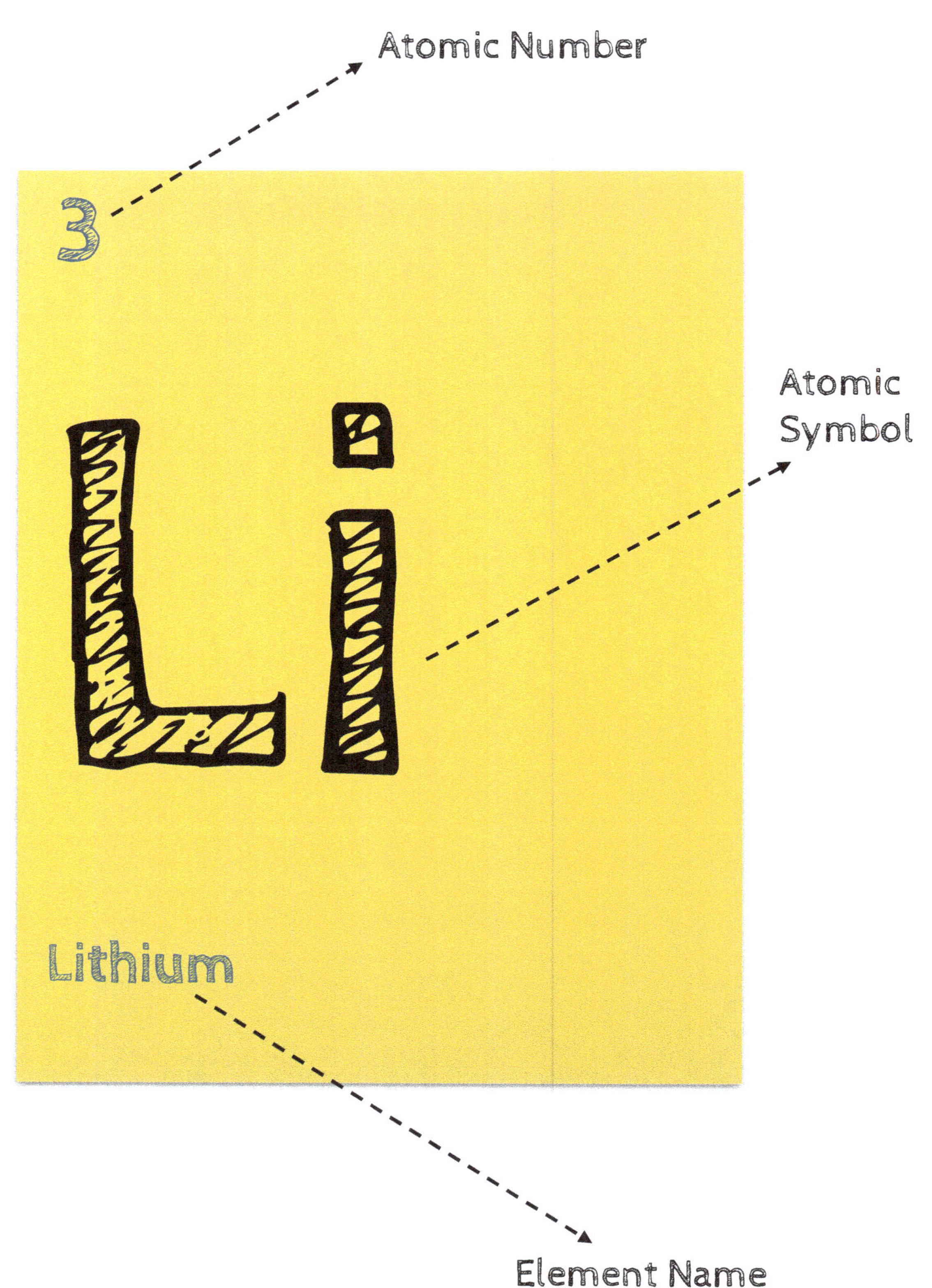

Atomic Number
3
Li
Atomic Symbol
Lithium
Element Name

FACTS ABOUT LITHIUM

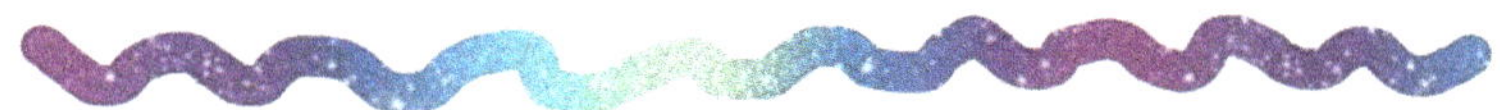

* Lithium is the lightest known metal

* Lithium is used in rechargeable batteries (such as the ones used in phones and computers)

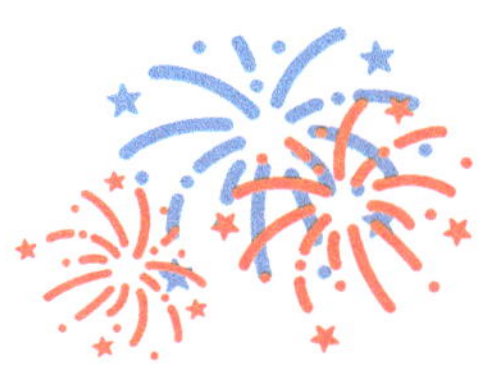

* When lithium is burned, it produces a bright red flame. Fireworks contain lithium to produce red sparks

(Emsley, 2001; Stwertka, 2002)

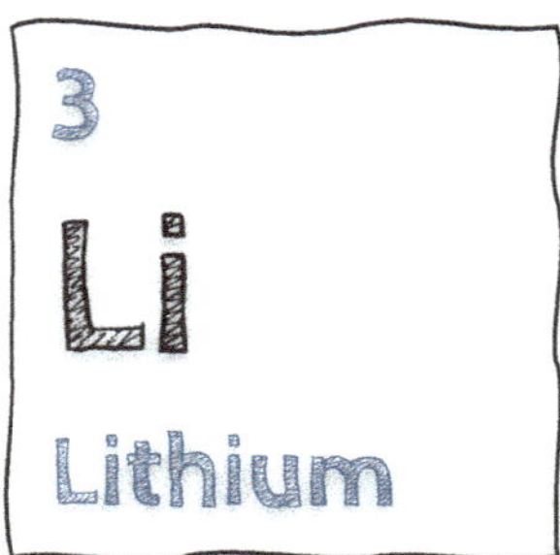

What item around you contains Lithium?

Draw it!

Write a short story that features the Lithium element.

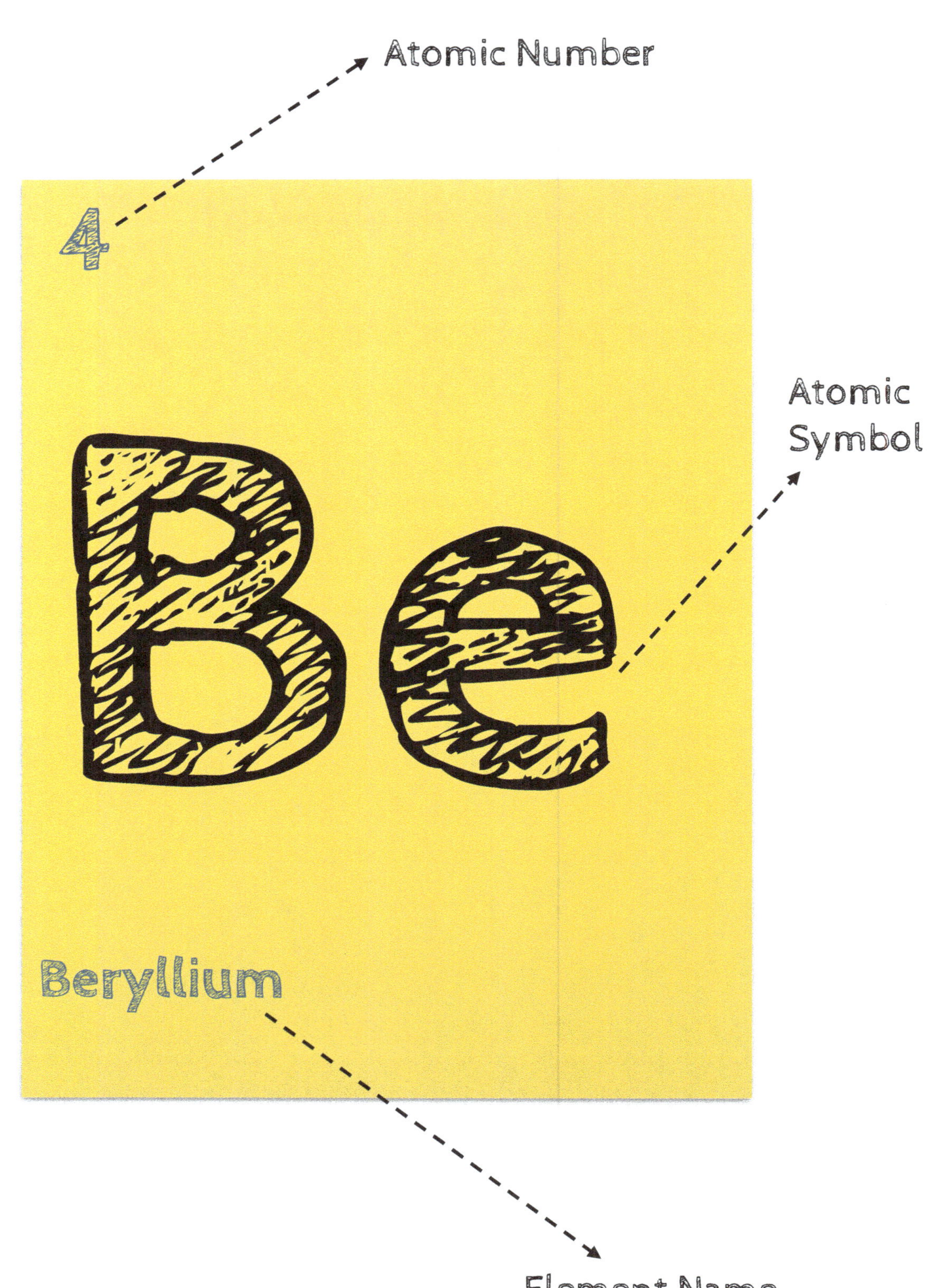

Atomic Number
4
Atomic Symbol
Be
Beryllium
Element Name

FACTS ABOUT BERYLLIUM

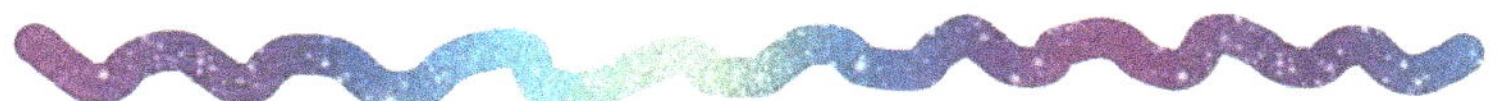

* The earth's crust contains a lot of beryllium. It is toxic to humans

* Pure beryllium has a white-gray metallic color

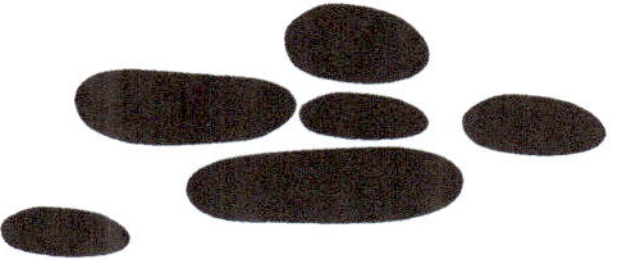

* Beryllium can be found in volcanic rocks, soil, and even the air

(Emsley, 2001; Stwertka, 2002)

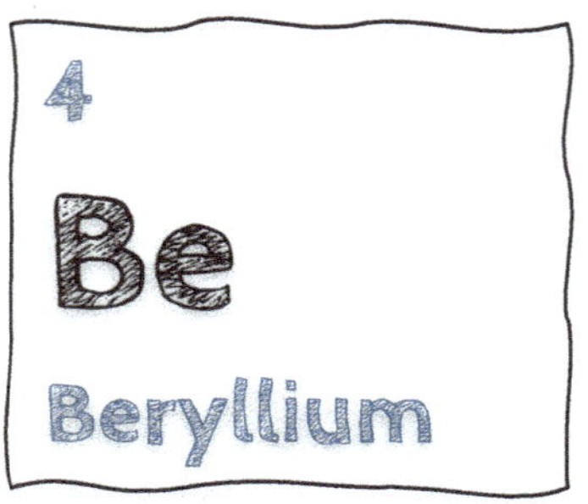

What item around you contains Beryllium?

Draw it!

Write a short story that features the
Lithium element.

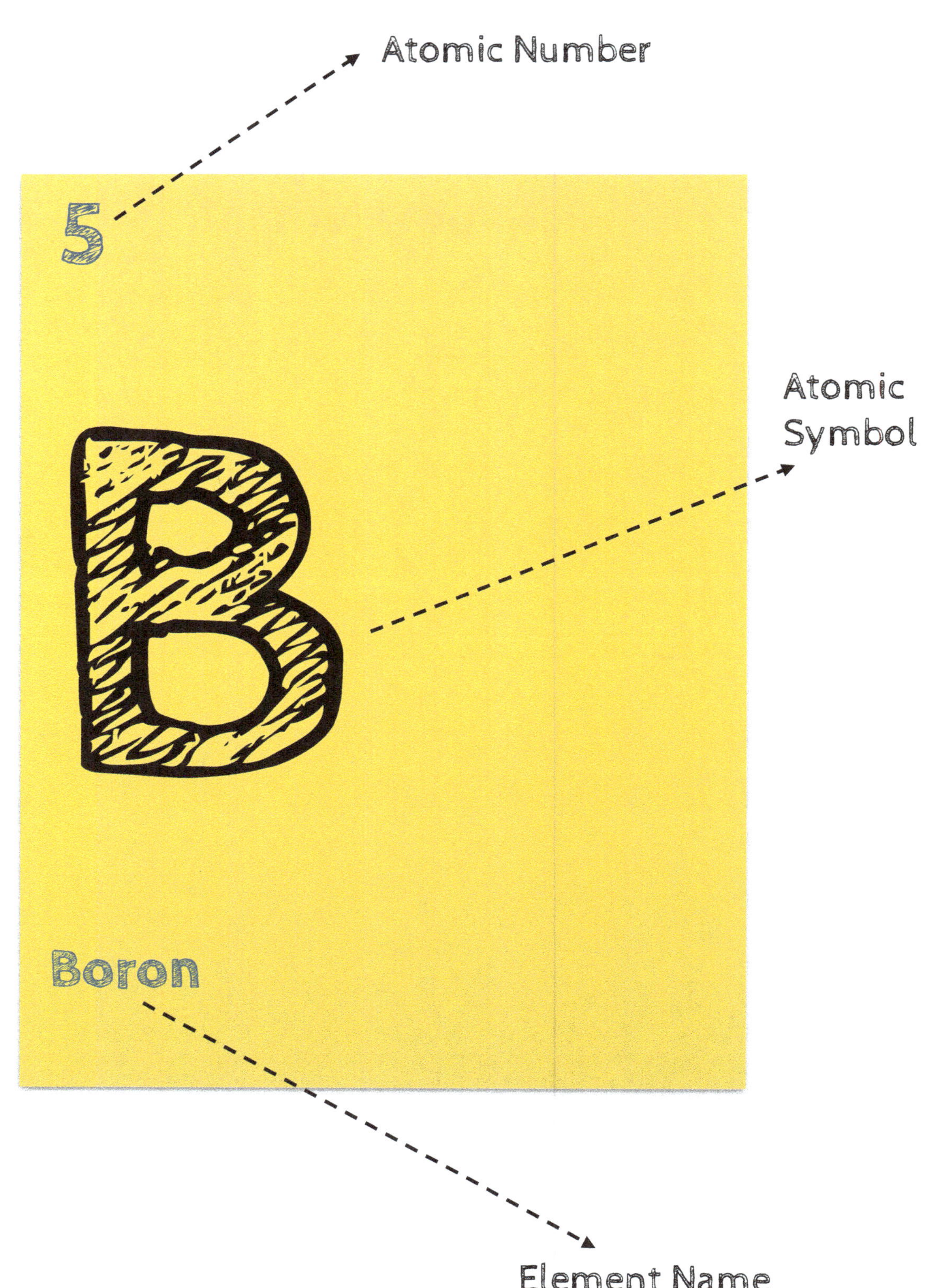

Atomic Number
5
B
Atomic Symbol
Boron
Element Name

FACTS ABOUT BORON

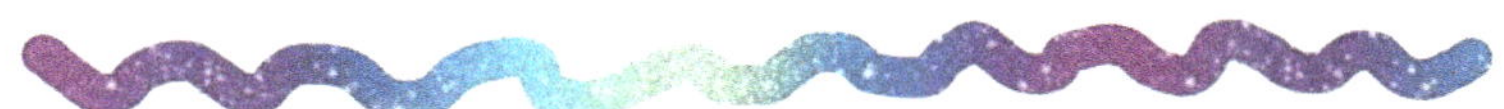

✳ Green plants need boron as a nutrient to survive. It is non-toxic to humans

✳ Boron is extremely hard and can resist heat very well

✳ Boron can be found in detergents, eyedrops, and agricultural fertilizers

(Emsley, 2001; Stwertka, 2002)

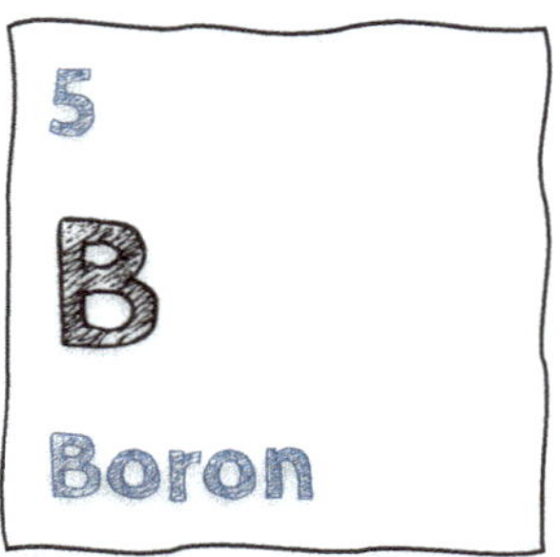

What item around you contains Boron?

Draw it!

Write a short story that features the
Boron element.

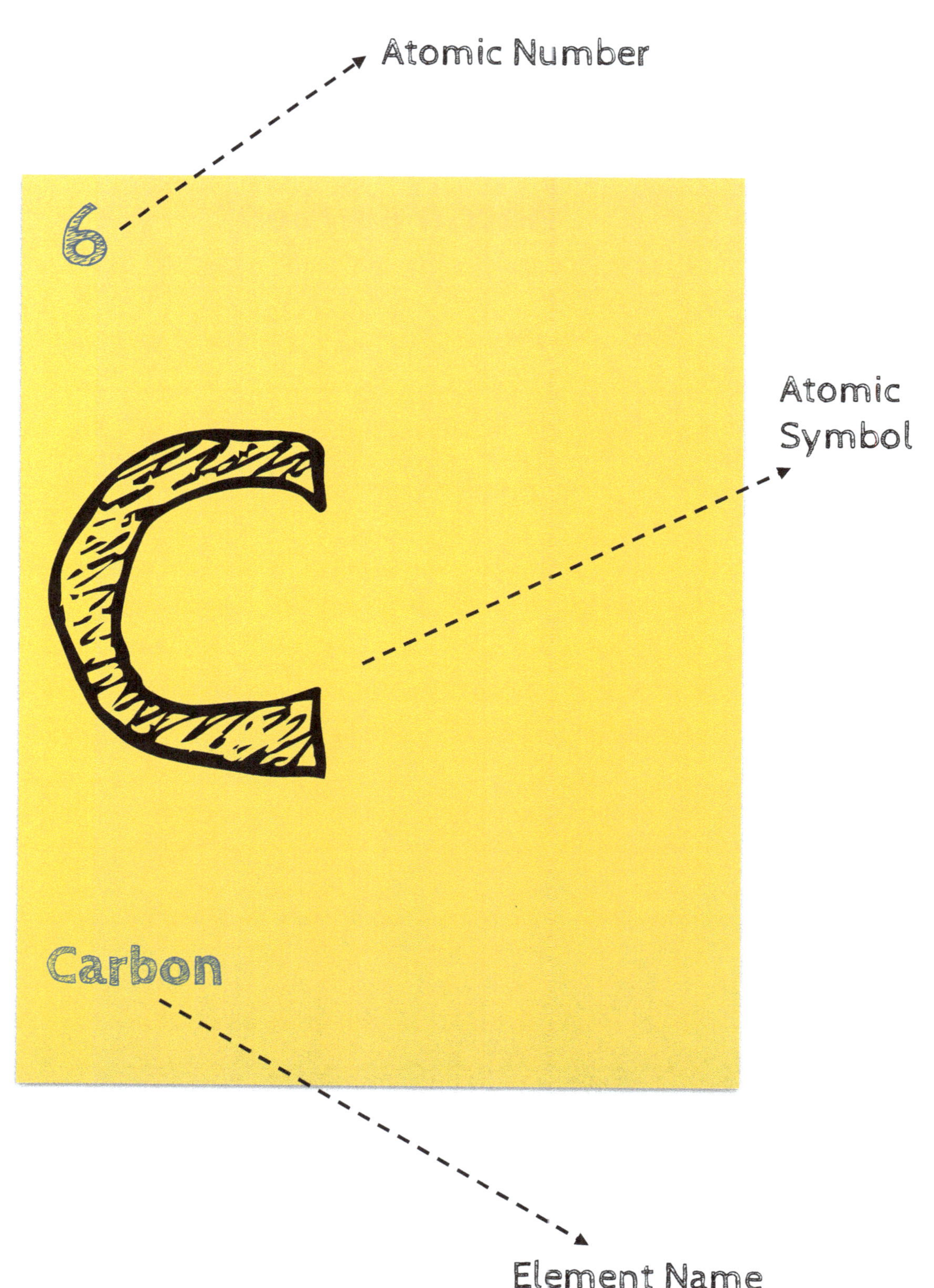
Atomic Number
6
Atomic Symbol
C
Carbon
Element Name

FACTS ABOUT CARBON

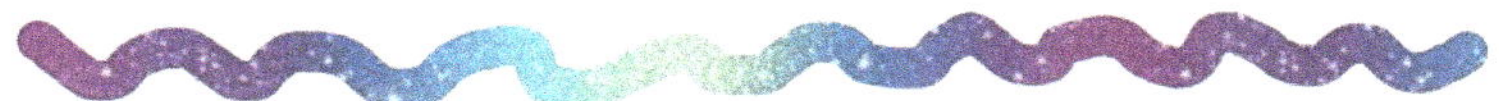

✵ Diamonds are made of pressurized carbon (one of the toughest materials on earth)

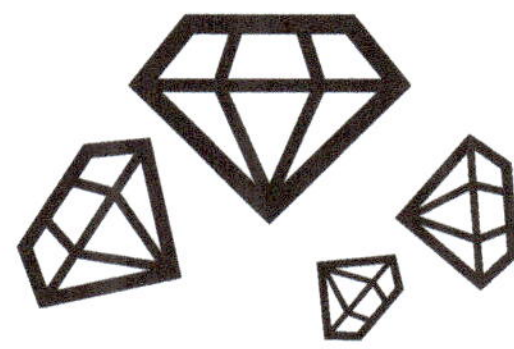

✵ Carbon is reactive. Graphite is a soft form of carbon used in pencils

✵ All living things contain carbon. When humans exhale, we breathe out carbon dioxide

(Emsley, 2001; Stwertka, 2002)

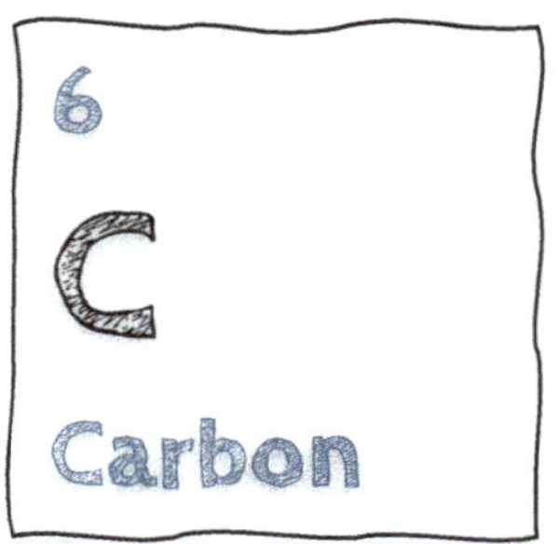

What item around you contains Carbon?

Draw it!

Write a short story that features the Carbon element.

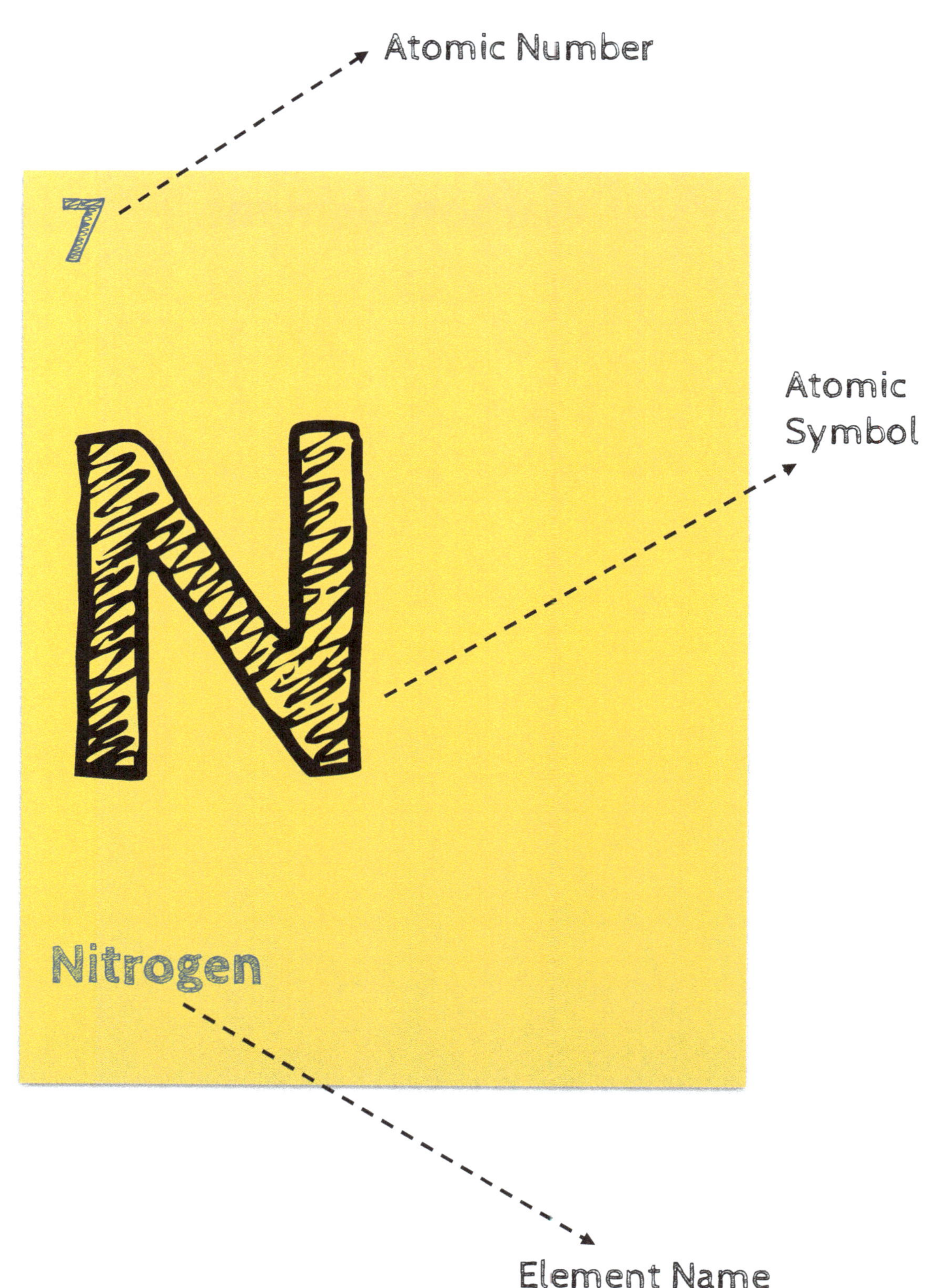

Atomic Number
7
N
Atomic
Symbol
Nitrogen
Element Name

FACTS ABOUT NITROGEN

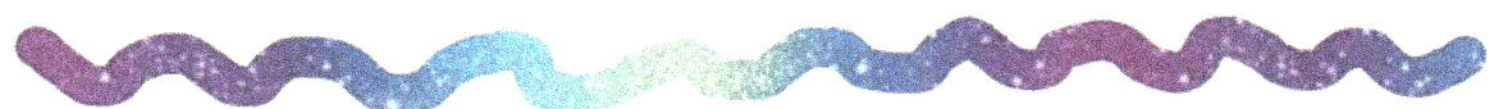

* Nitrogen has no smell, color, or taste

* Most of the earth's atmosphere is made up of nitrogen gas (78.1%)

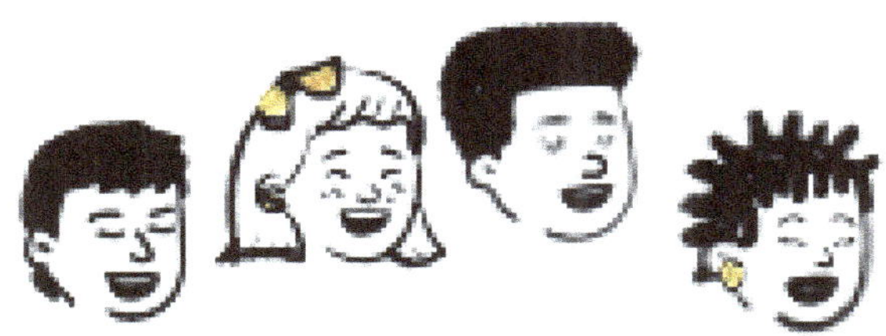

* Nitrogen is used in laughing gas to reduce pain, and refrigeration to preserve food freshness

(Emsley, 2001; Stwertka, 2002)

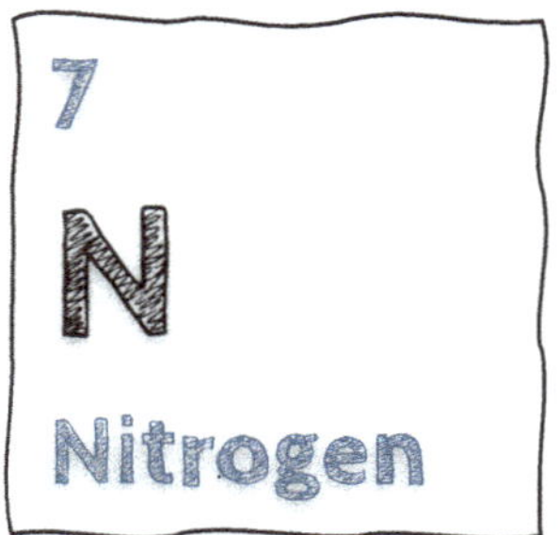

What item around you contains Nitrogen?

Draw it!

Write a short story that features the
Nitrogen element.

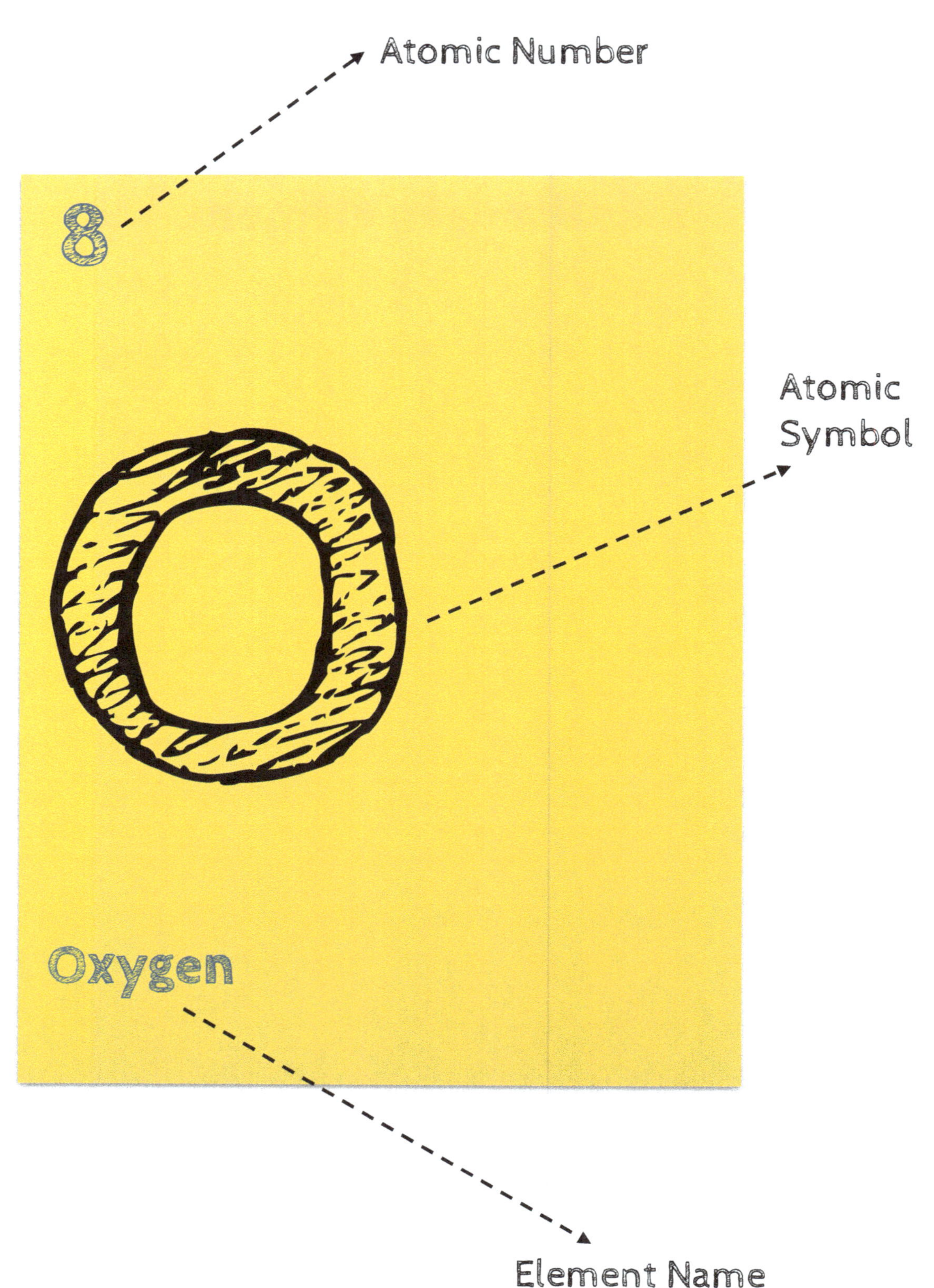

Atomic Number

Atomic Symbol

Element Name

FACTS ABOUT OXYGEN

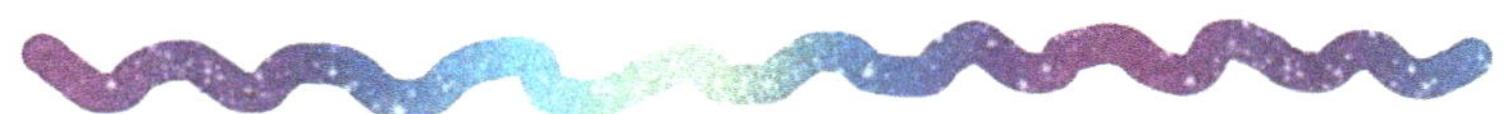

* Oxygen has no smell, color, or taste. It is needed to support life on earth

* Oxygen binds to hydrogen to make water

* Oxygen makes up 21% of our atmosphere on earth

(Emsley, 2001; Stwertka, 2002)

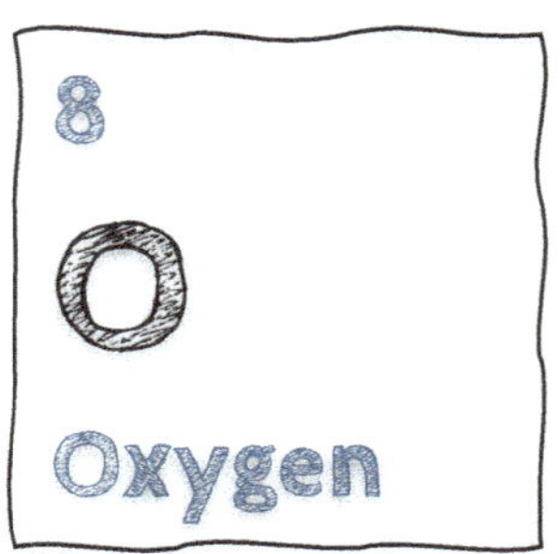

What item around you contains Lithium?

Draw it!

Write a short story that features the
Oxygen element.

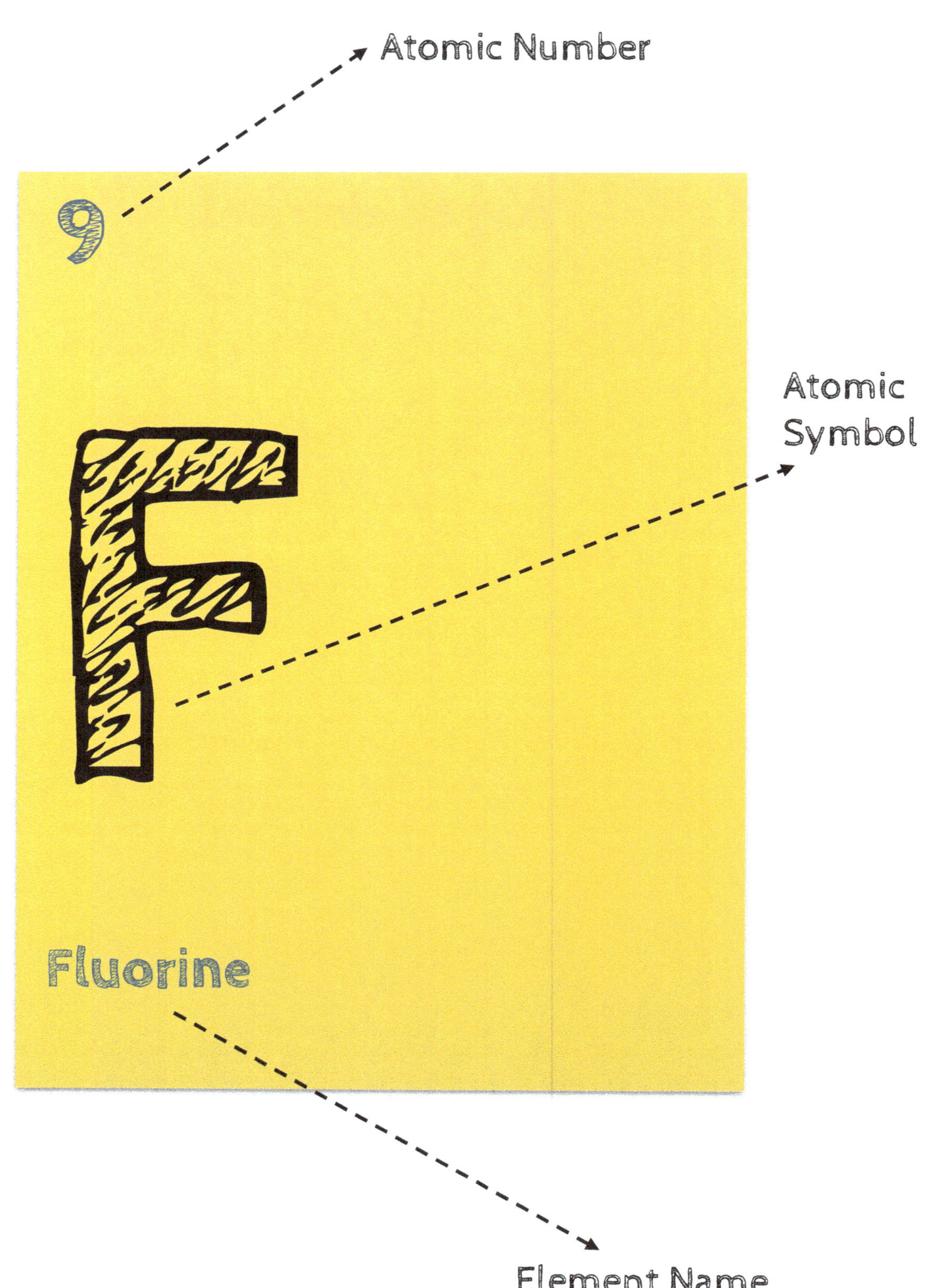
Atomic Number
9
Atomic
Symbol
F
Fluorine
Element Name

FACTS ABOUT FLUORINE

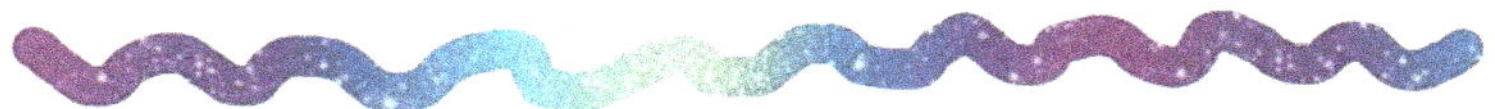

* Fluorine is the most reactive of off the non-metals

* Fluorine is used to make Teflon, which is used in frying pans, ad other cookware

* Fluorine is used on the toothpaste that you use to brush your teeth

(Emsley, 2001; Stwertka, 2002)

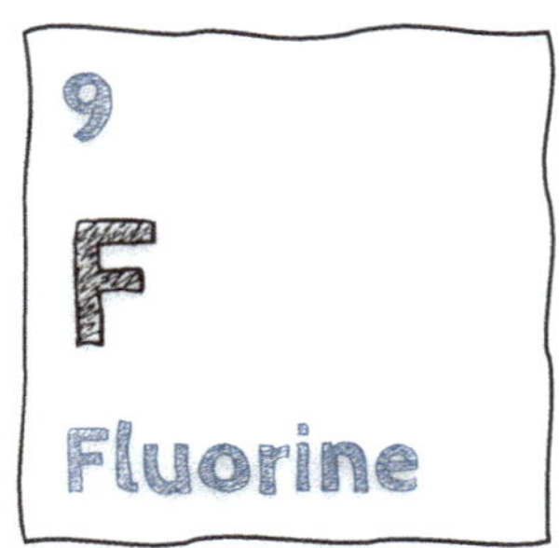

What item around you contains Fluorine?

Draw it!

STORY TIME...

Write a short story that features the
Fluorine element.

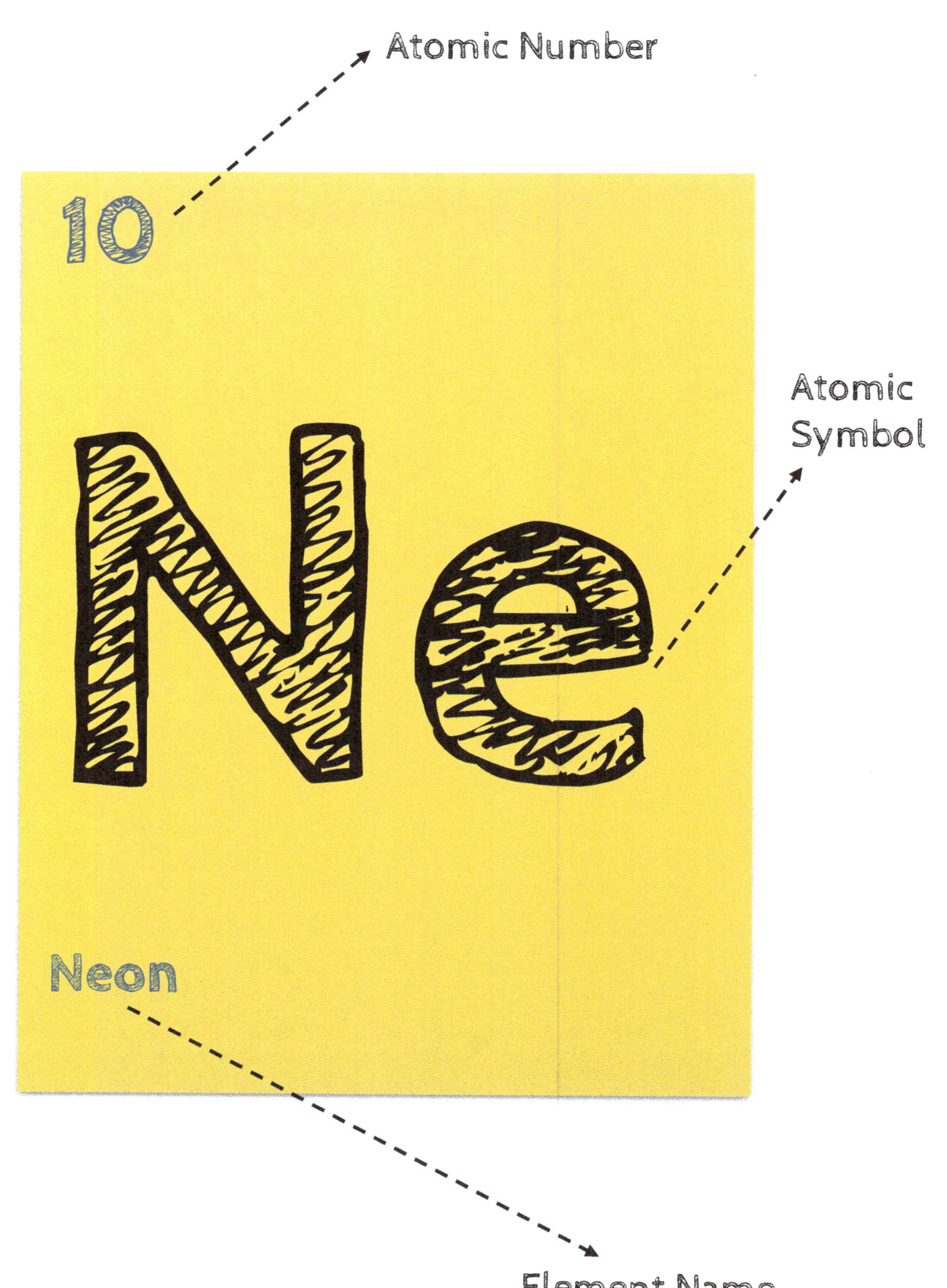

Atomic Number
10
Ne
Atomic Symbol
Neon
Element Name

FACTS ABOUT NEON

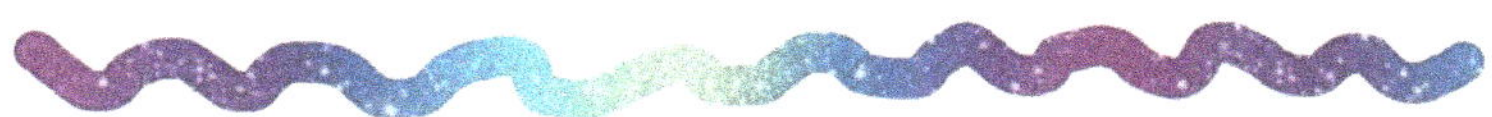

✻ Neon is a gas that does not have any color or smell (colorless and odorless)

✻ The Earth's atmosphere has a small amount of neon in it

✻ Neon is used in the neon light signs that you may see at restaurants

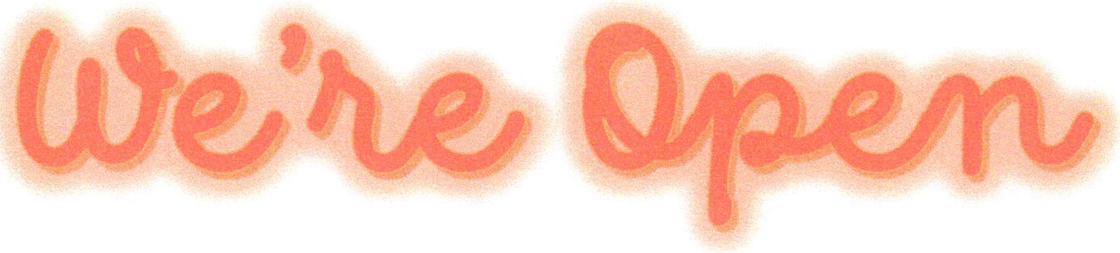

(Emsley, 2001; Stwertka, 2002)

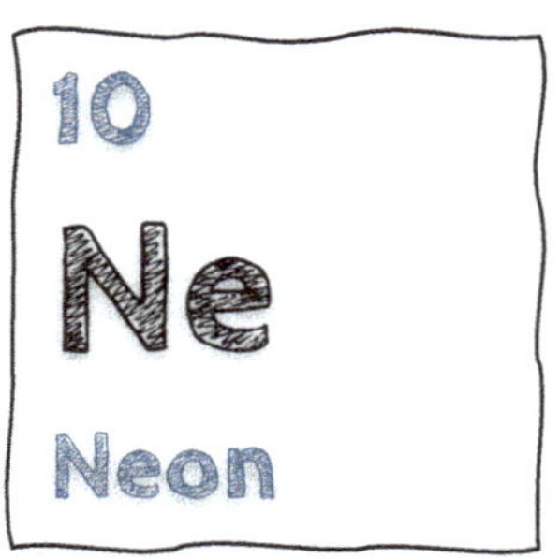

What item around you contains Neon?

Draw it!

STORY TIME...

Write a short story that features the Neon element.

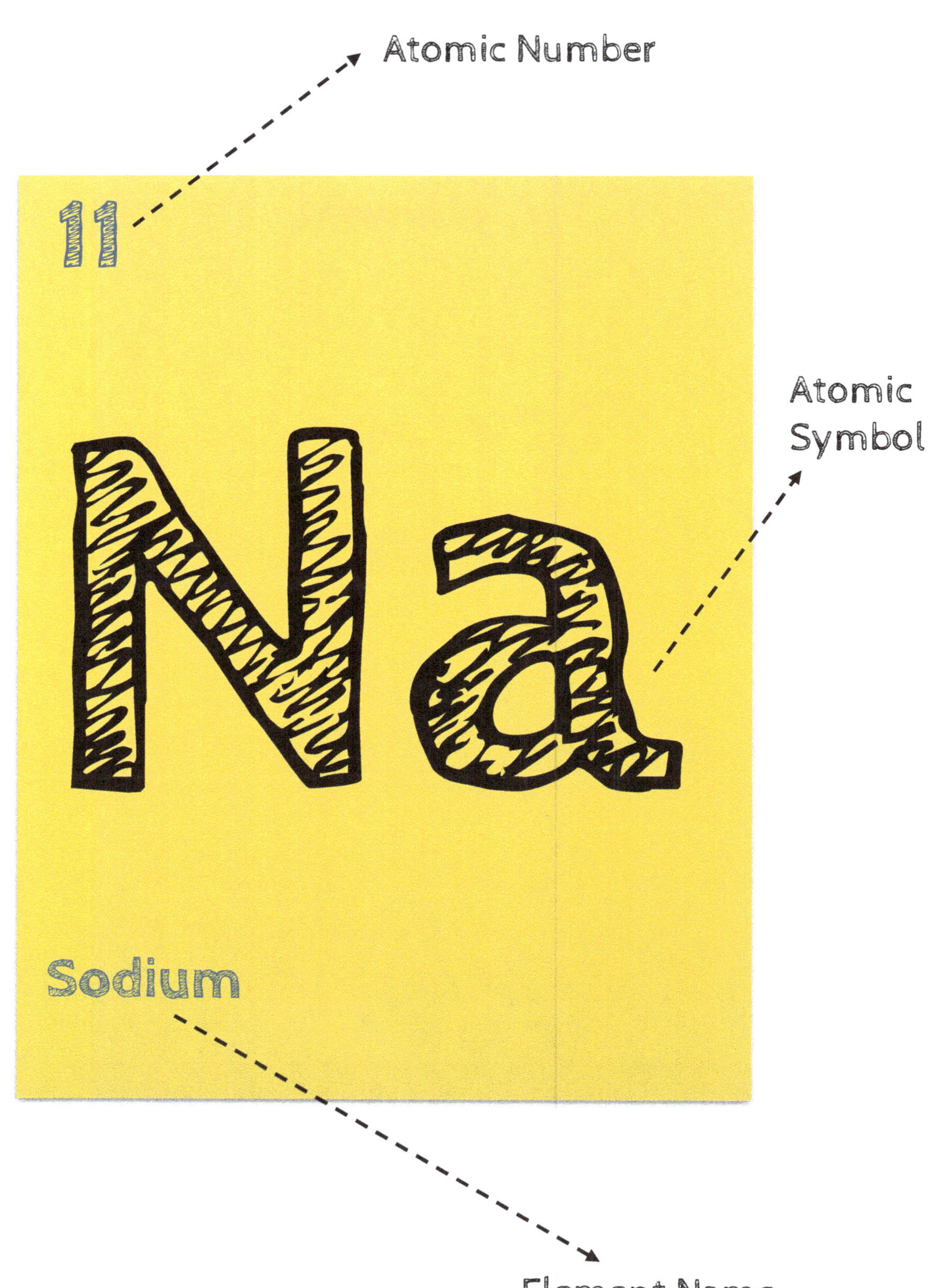

Atomic Number
11
Atomic Symbol
Na
Sodium
Element Name

FACTS ABOUT SODIUM

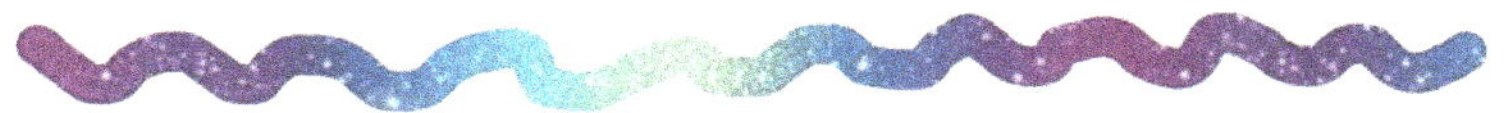

* Sodium is a soft metal that is silvery-white in color

* Sodium is used to make salt

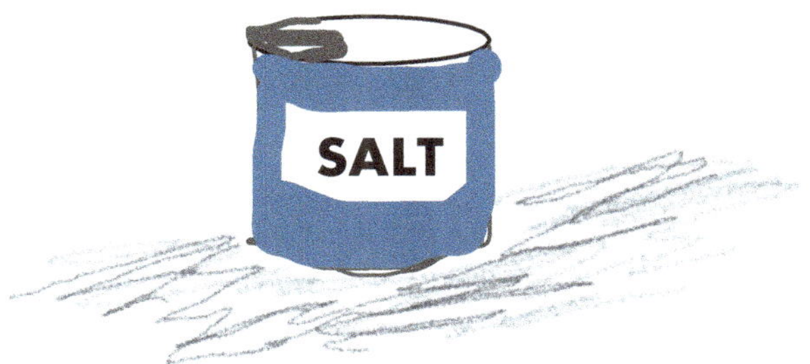

* Sodium can be found in soaps, laundry detergents, and dyes

(Emsley, 2001; Stwertka, 2002)

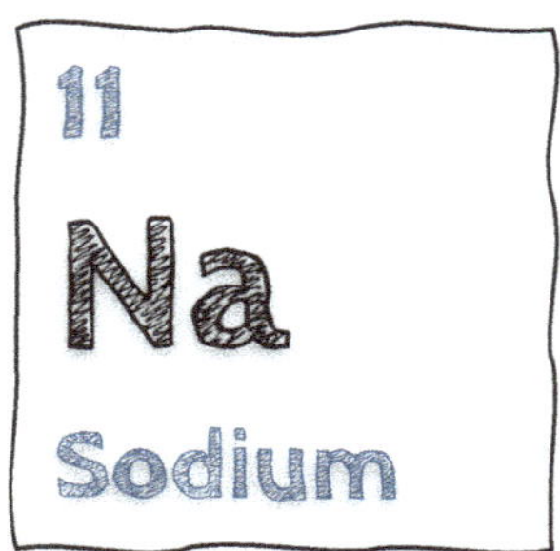

What item around you contains Sodium?

Draw it!

Write a short story that features the
Sodium element.

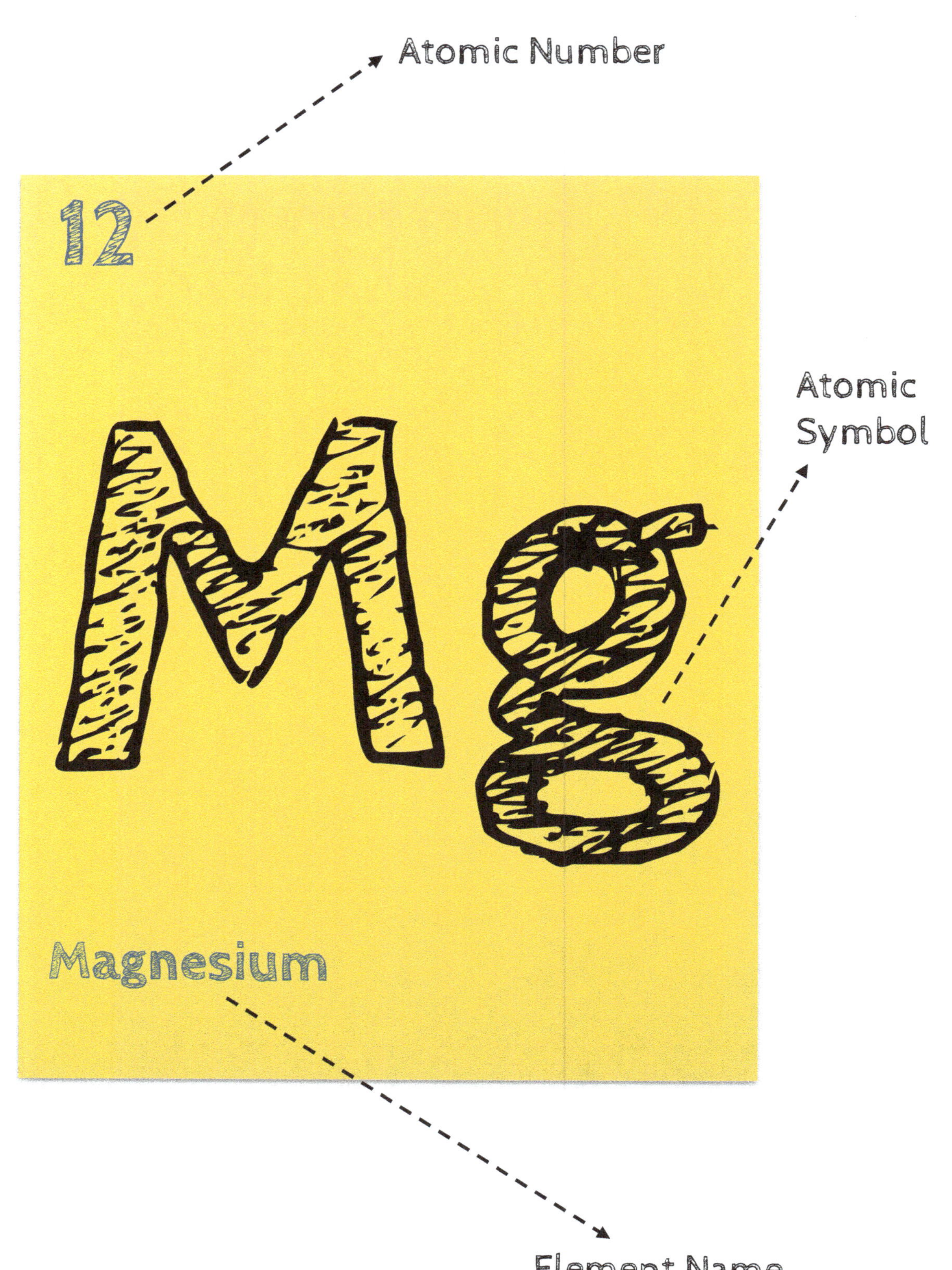

Atomic Number
12
Mg
Magnesium
Atomic Symbol
Element Name

FACTS ABOUT MAGNESIUM

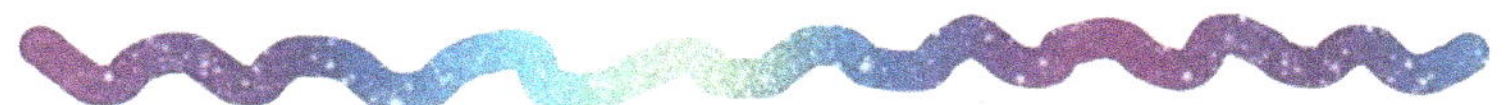

* Magnesium is used to make flares and fireworks

* Magnesium is important for our muscles to work and produce energy in our human bodies

* Chlorophyll is the compound that gives green plants their green color, and it contains Magnesium

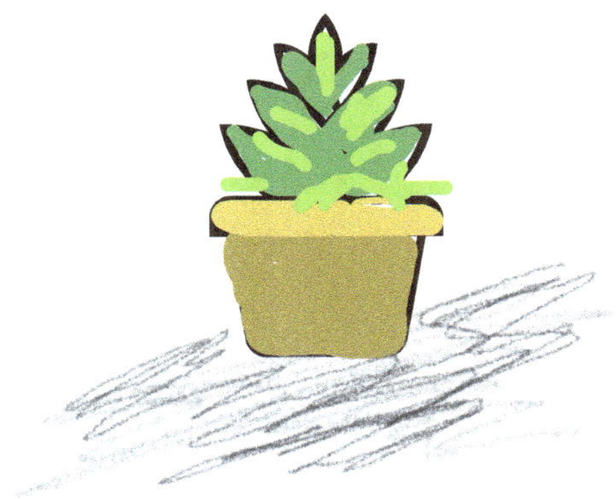

(Emsley, 2001; Stwertka, 2002)

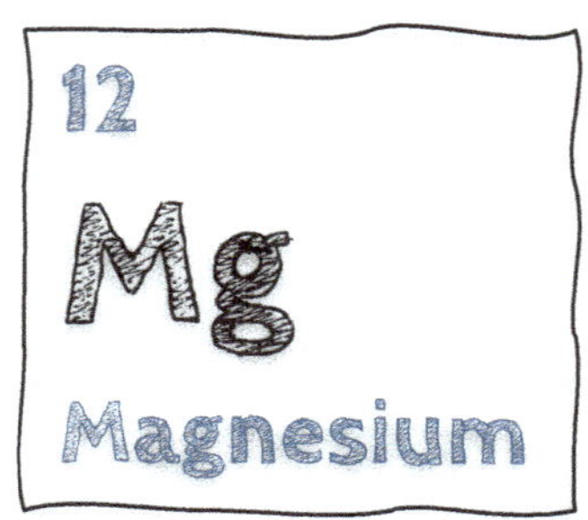

What item around you contains Magnesium?

Draw it!

Write a short story that features the
Magnesium element.

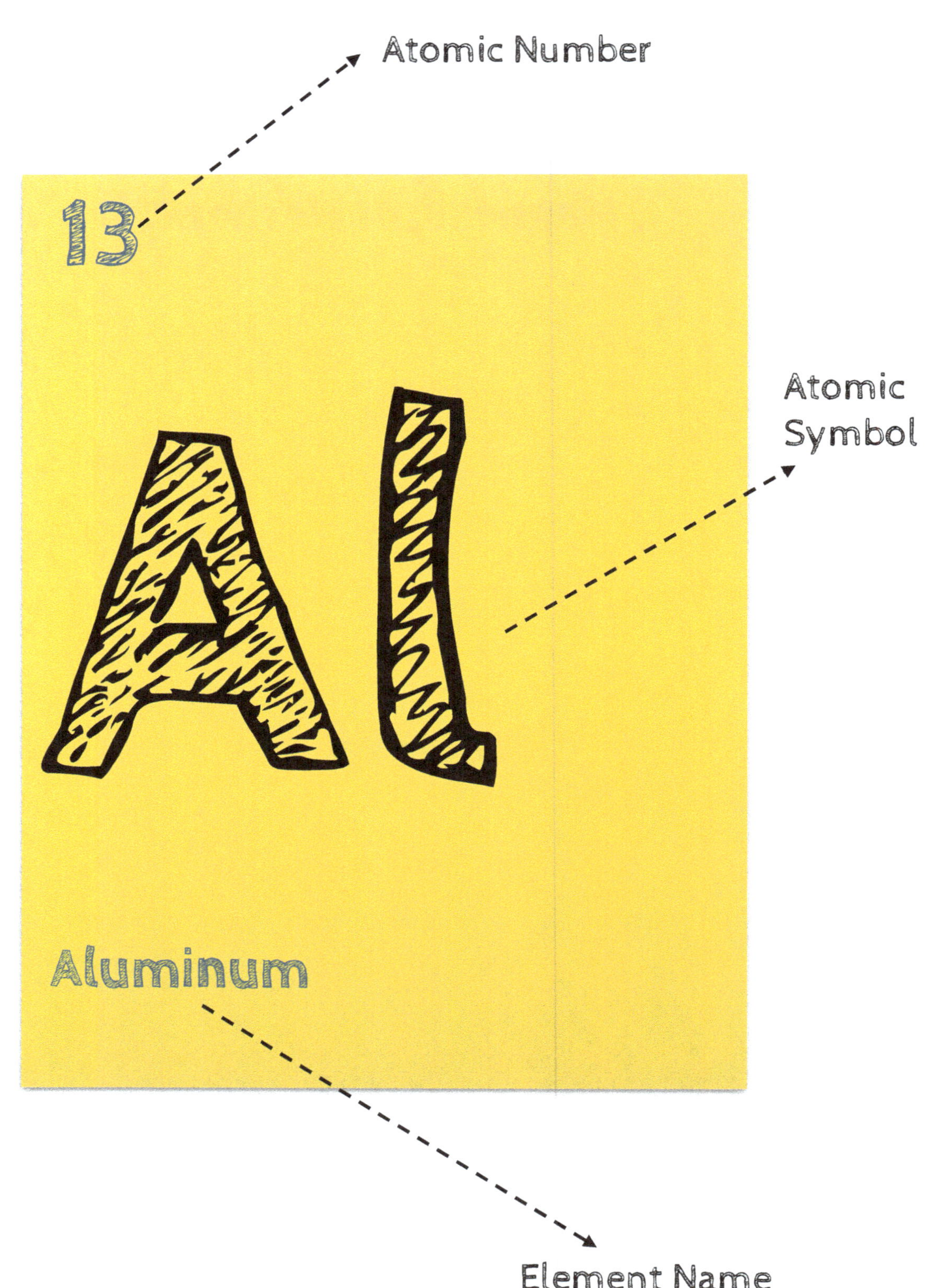

Atomic Number

Atomic Symbol

Element Name

FACTS ABOUT ALUMINUM

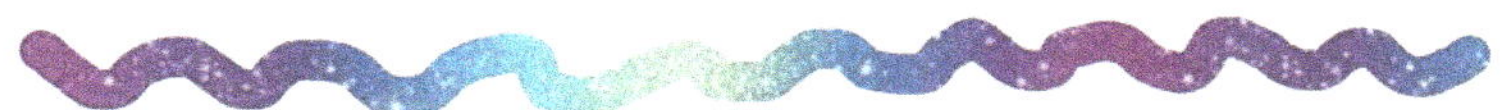

* Aluminum is a lightweight metal

* Aluminum is used to make cans. cooking foil, and some window frames

* Airplane and car engines are also made of aluminum

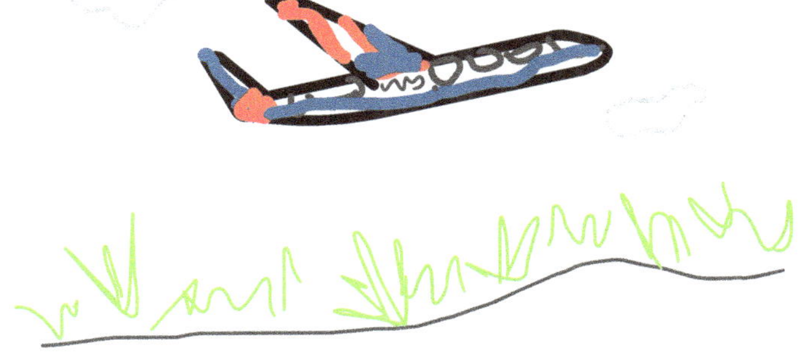

(Emsley, 2001; Stwertka, 2002)

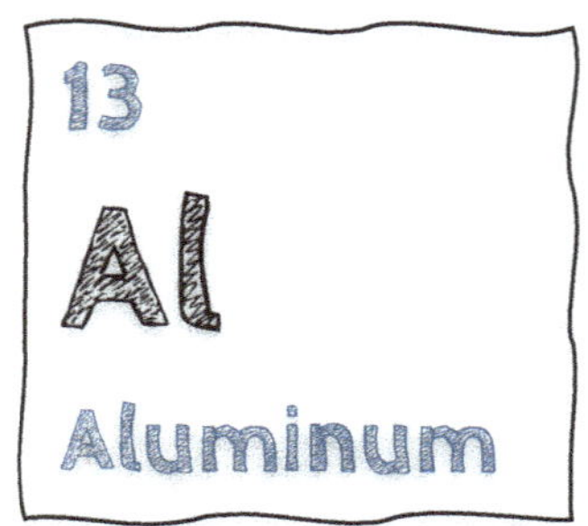

What item around you contains Aluminum?

Draw it!

STORY TIME...

Write a short story that features the
Aluminum element.

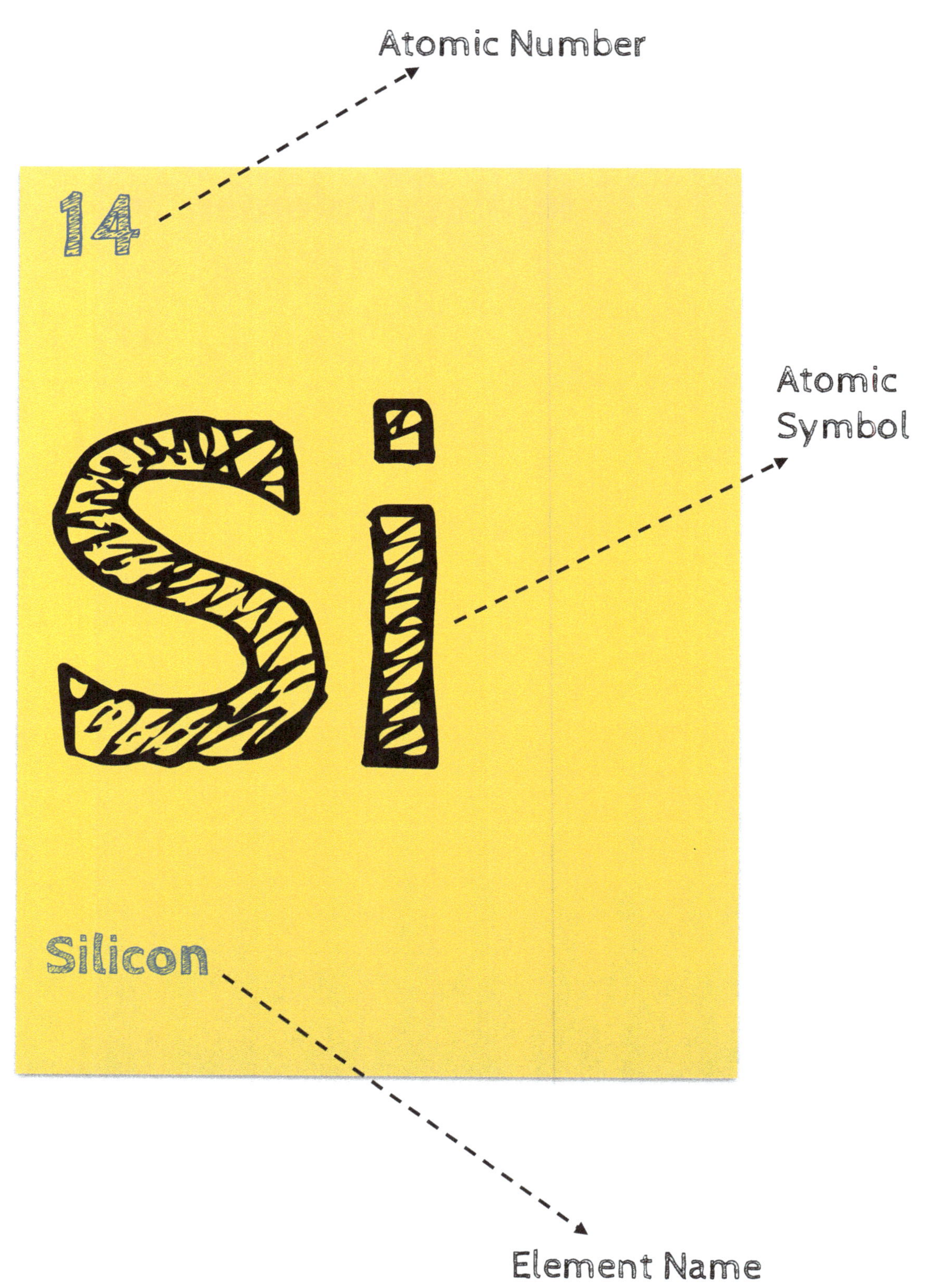

Atomic Number
14
Si
Silicon
Atomic Symbol
Element Name

FACTS ABOUT SILICON

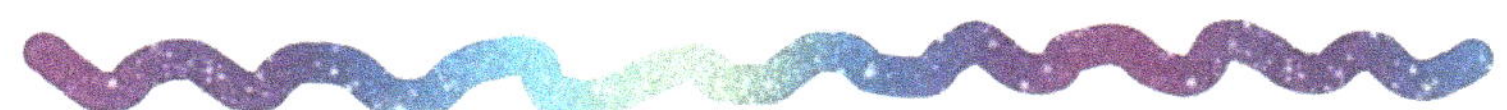

* Silicon is hard, and has a blue-gray metallic color

* Silicon is used to make glass, ceramics, and solar cells

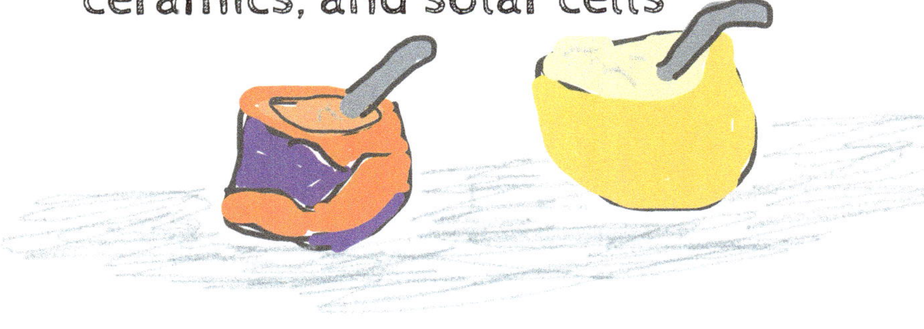

* Silicon is used in semiconductors, which are found in electronics like cell phones, refrigerators, televisions, and even washing machines

(Emsley, 2001; Stwertka, 2002)

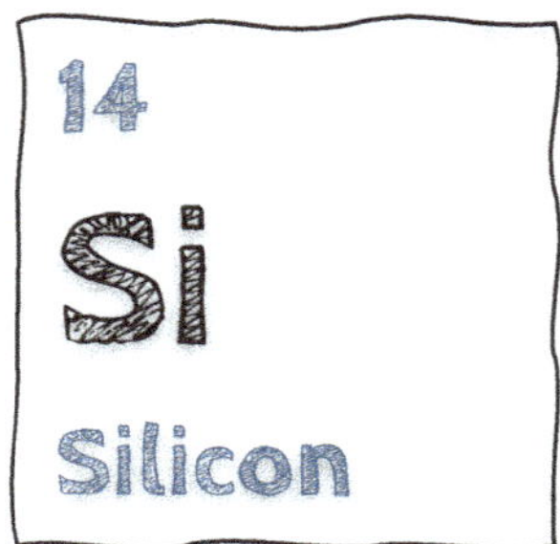

What item around you contains Silicon?

Draw it!

Write a short story that features the
Silicon element.

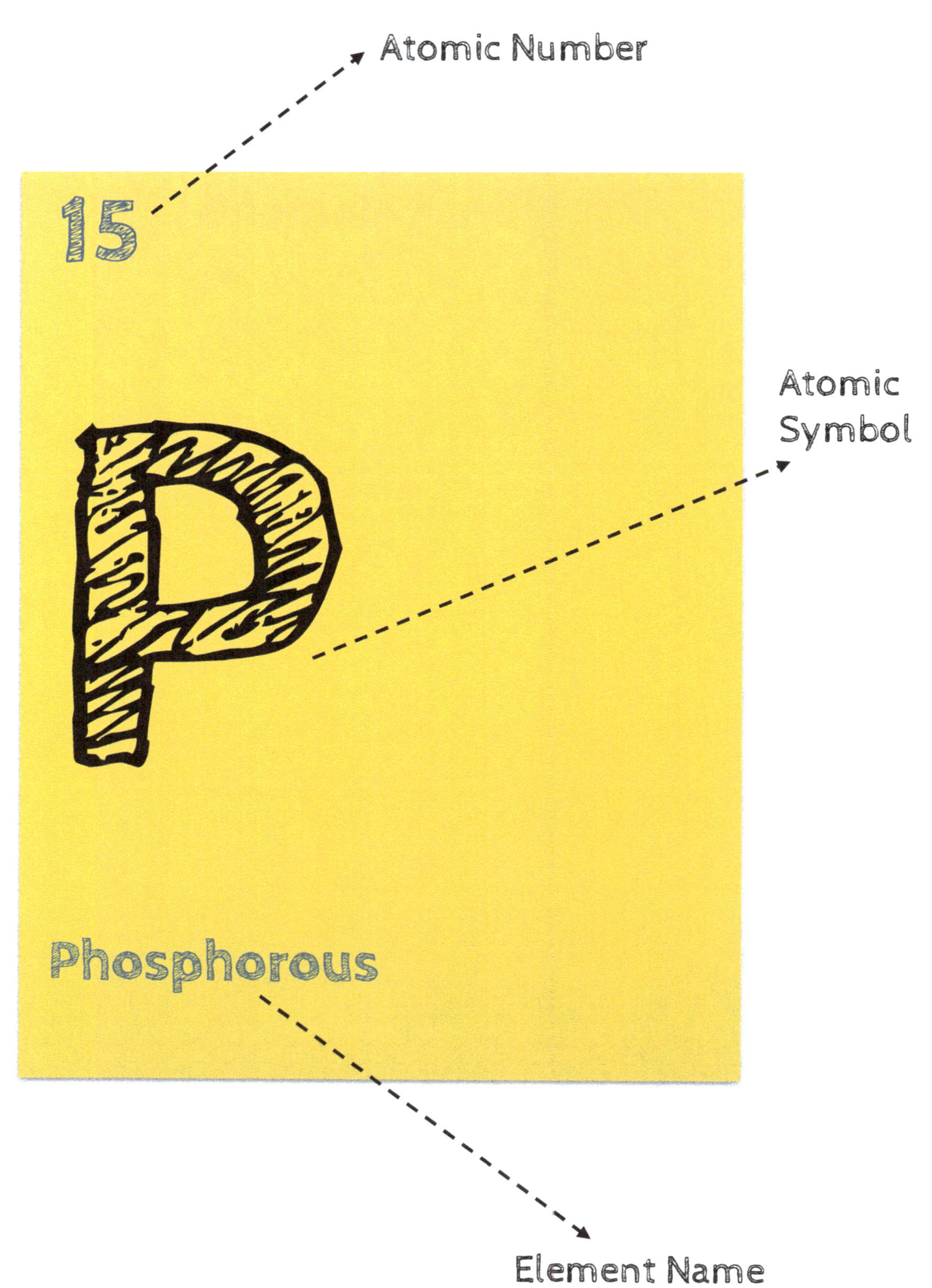
Atomic Number
15
P
Atomic
Symbol
Phosphorous
Element Name

FACTS ABOUT PHOSPHOROUS

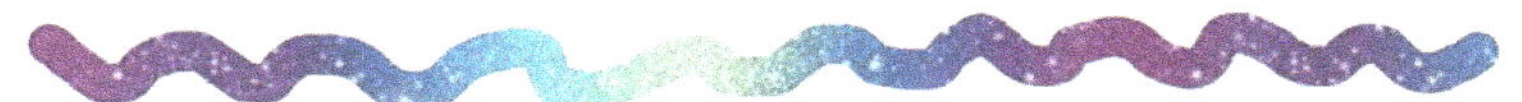

* Phosphorous is white in color, when it is in its pure form

* Phosphorous is used to make fertilizers to help plants grow

* Phosphorous is found in the human body. It's in our DNA and RNA

(Emsley, 2001; Stwertka, 2002)

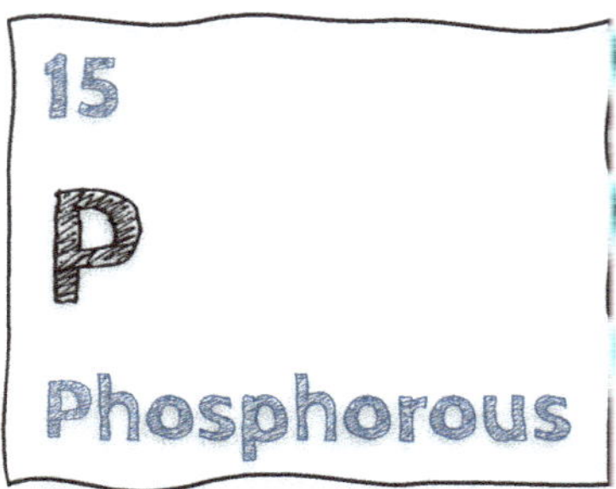

What item around you contains Phosphorous?

Draw it!

Write a short story that features the
Phosphorous element.

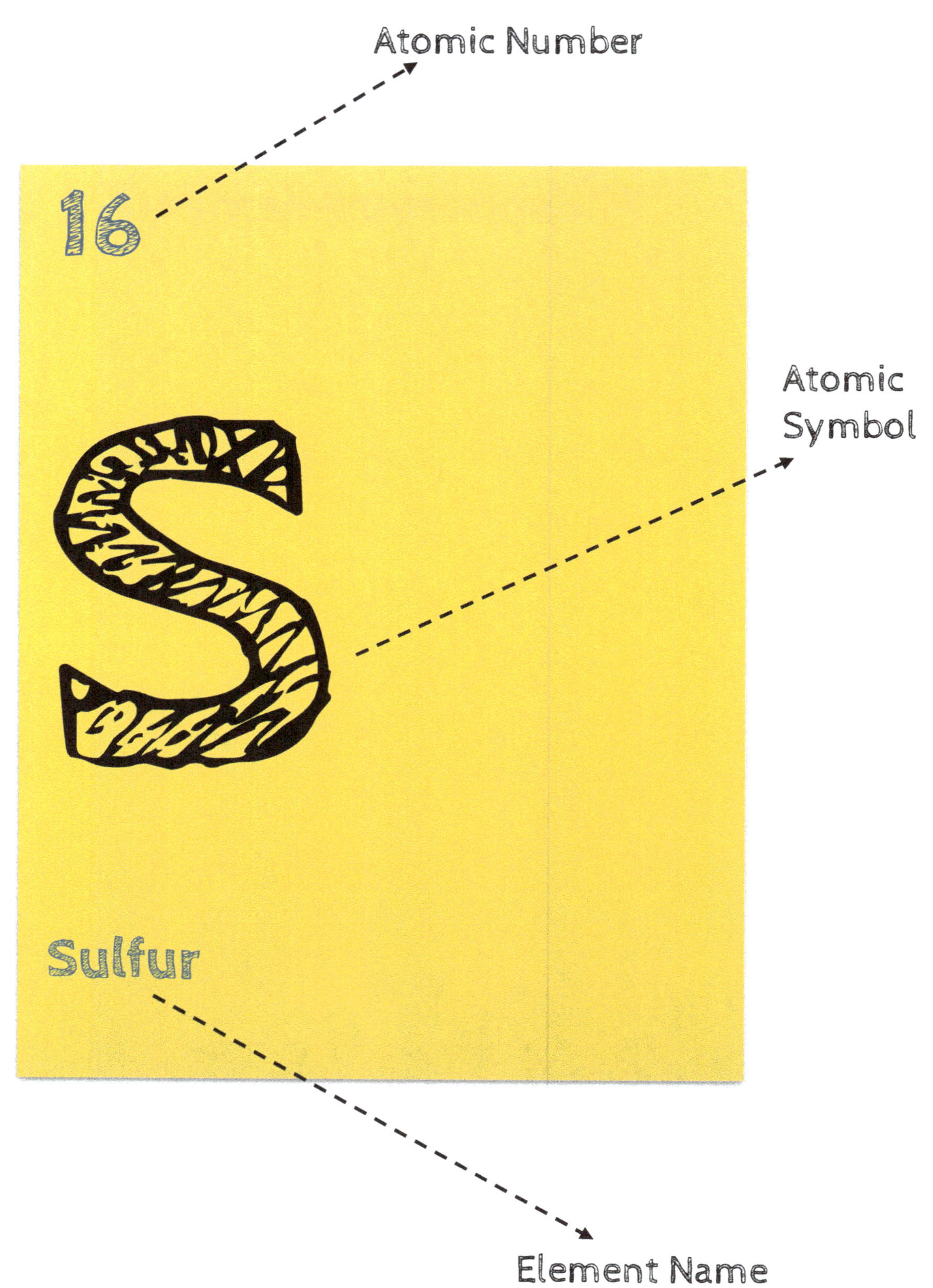

16
S
Sulfur
Atomic Number
Atomic Symbol
Element Name

FACTS ABOUT SULFUR

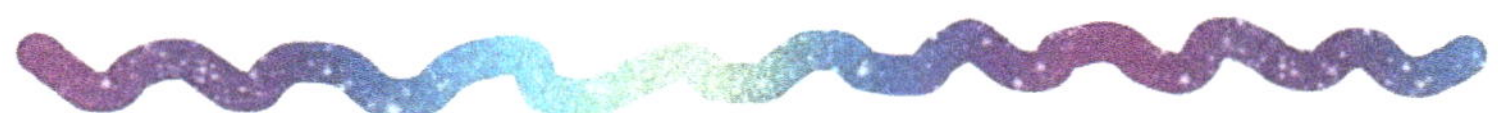

❋ Sulfur has a light-yellow color in its solid form, but is amber in color when it is melted

❋ In its pure state, sulfur has no odor, but it may also have a light rotten egg smell

❋ Sulfur is used to make sulfuric acid, detergents, and medicine to get rid of fungi

(Emsley, 2001; Stwertka, 2002)

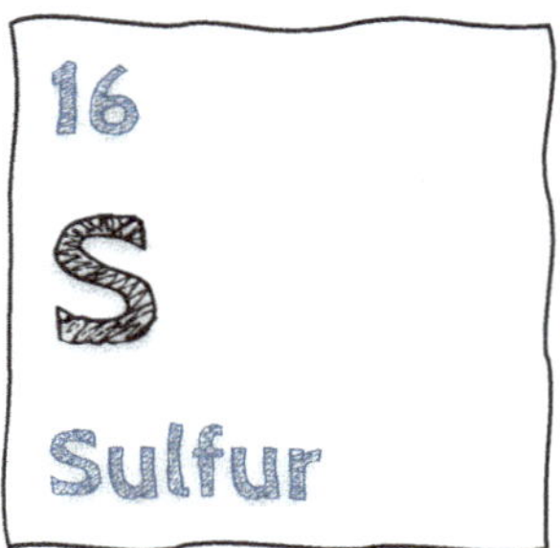

What item around you contains Sulfur?

Draw it!

Write a short story that features the
Sulfur element.

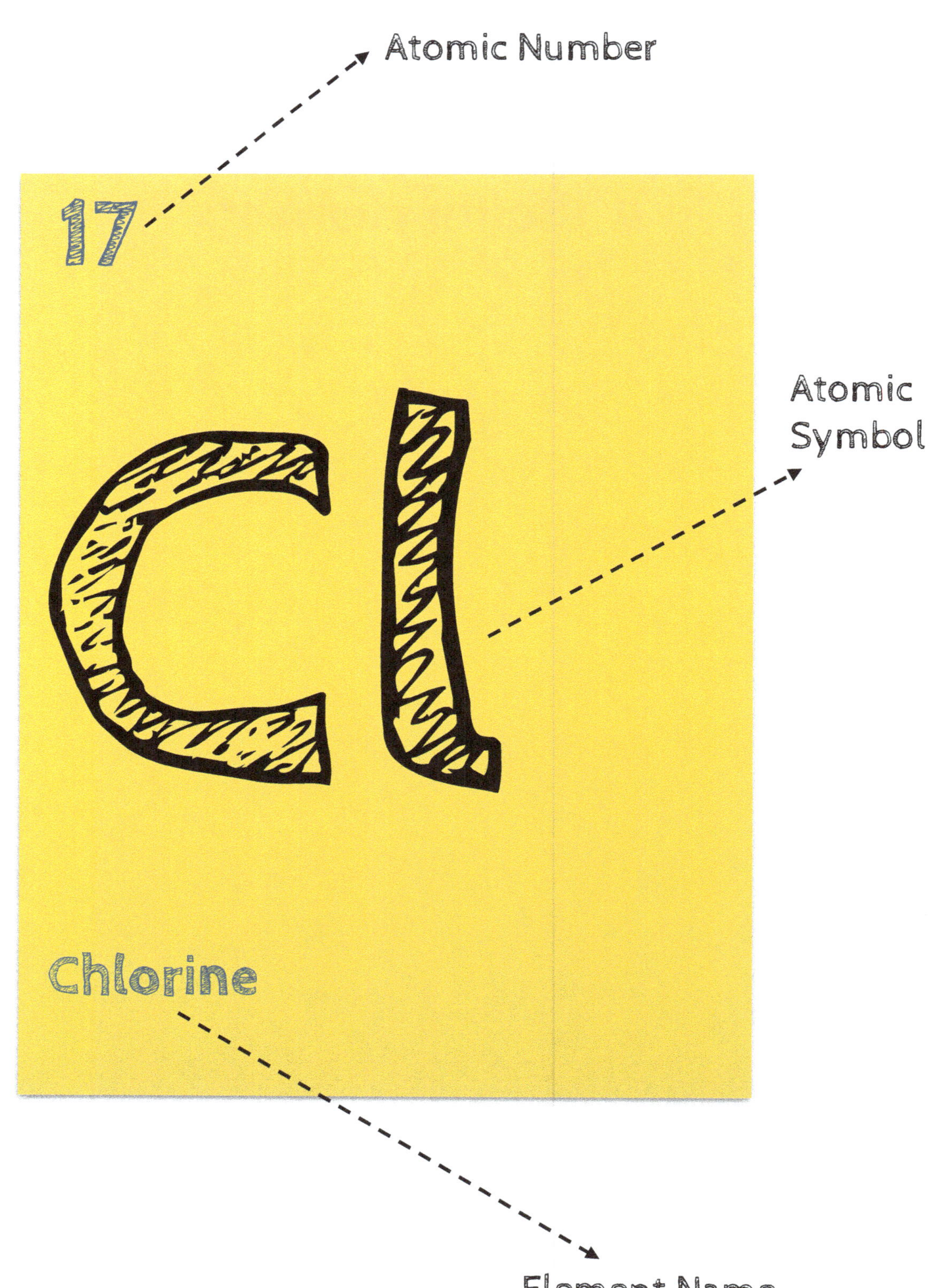
Atomic Number
17
Atomic Symbol
cl
Chlorine
Element Name

FACTS ABOUT CHLORINE

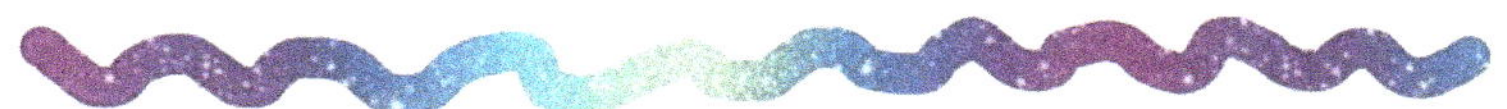

⚛ Chlorine is a gas that is yellow-green in color

⚛ Chlorine has a very strong smell, and is used to kill germs in swimming pools

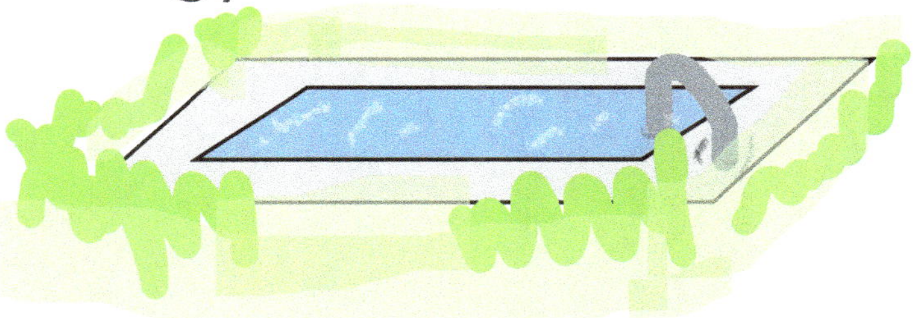

⚛ Bleach and tear gas contain chlorine

(Emsley, 2001; Stwertka, 2002)

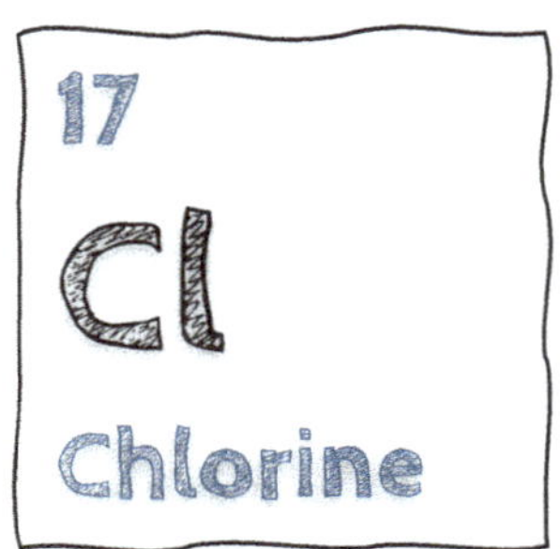

What item around you contains Chlorine?

Draw it!

Write a short story that features the
Chlorine element.

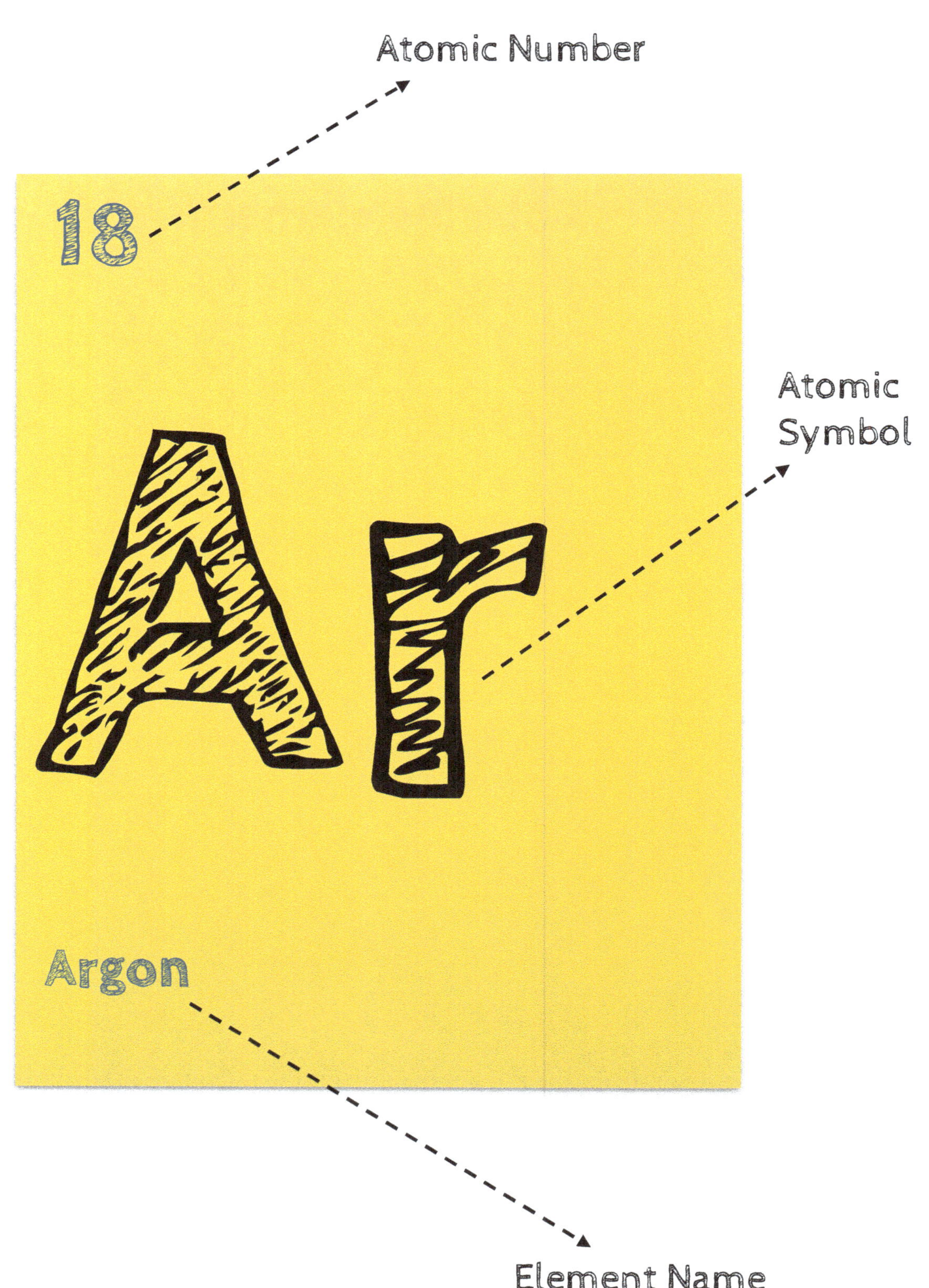
Atomic Number
18
Atomic
Symbol
Ar
Argon
Element Name

FACTS ABOUT ARGON

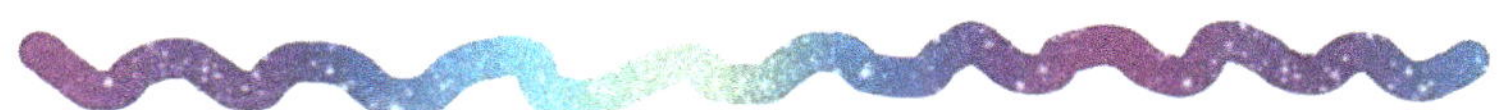

* Argon is a colorless and odorless gas

* Argon is used in fluorescent light bulbs

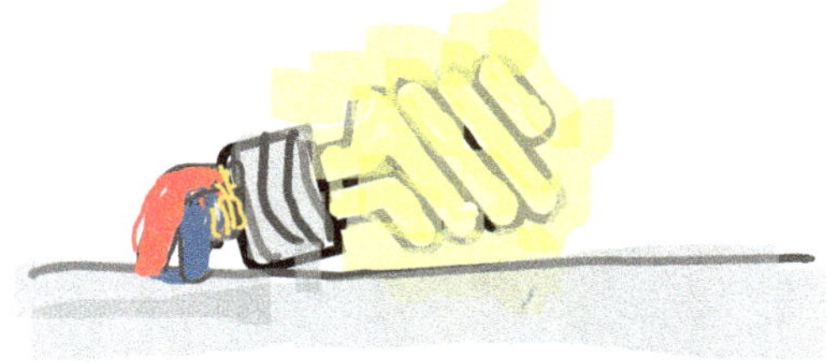

* Argon can be used as a gas shield in the process of welding metal

(Emsley, 2001; Stwertka, 2002)

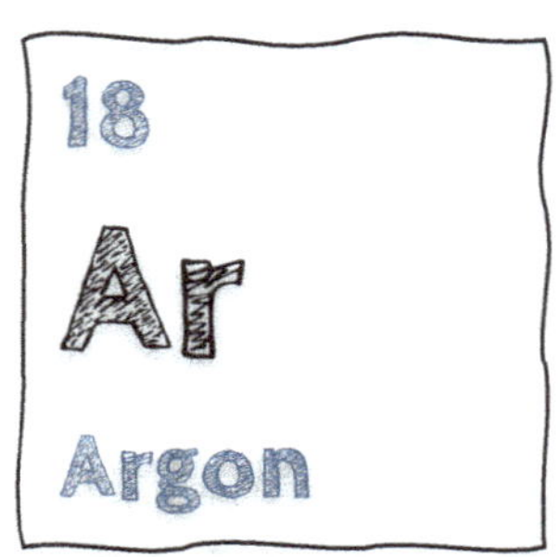

What item around you contains Argon?

Draw it!

Write a short story that features the
Argon element.

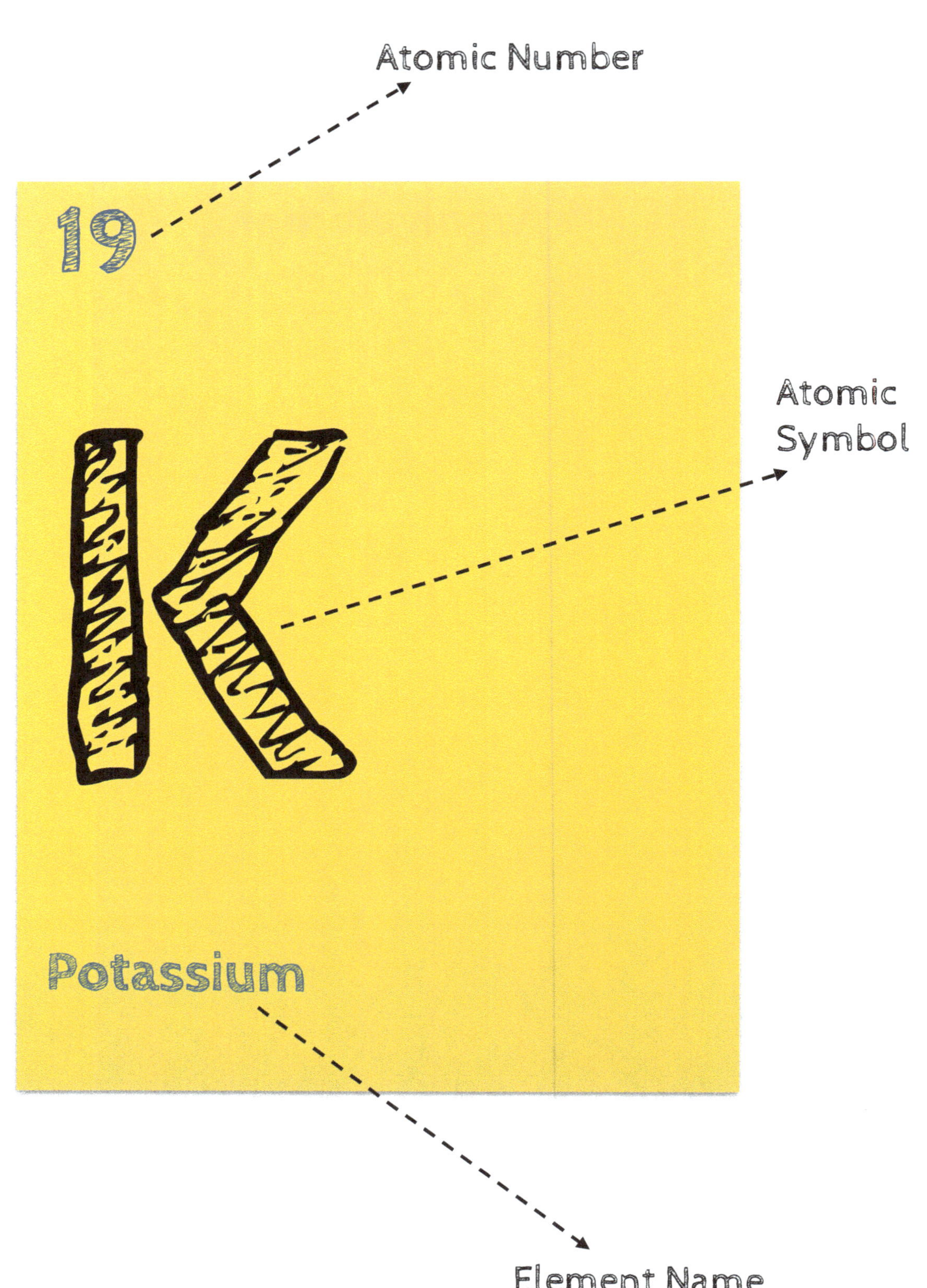
Atomic Number
19
Atomic
Symbol
K
Potassium
Element Name

FACTS ABOUT POTASSIUM

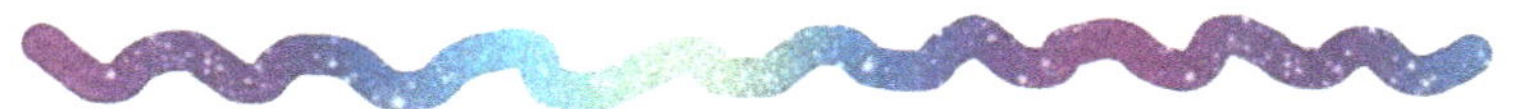

❋Potassium is a metal that is soft and white, with a silvery sheen

❋Potassium reacts very strongly, when it touches water

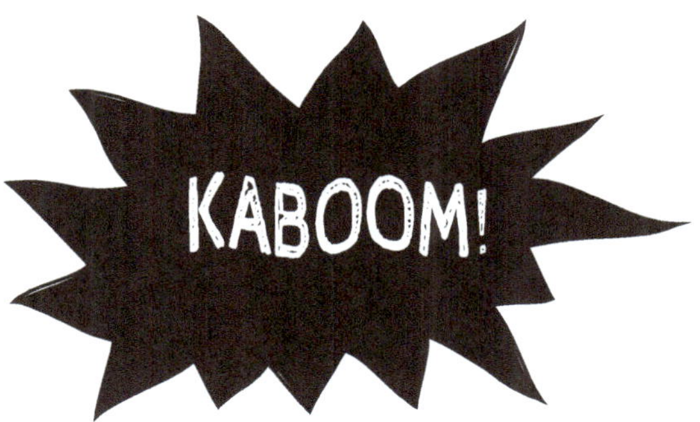

❋Potassium is used to make soaps. It also helps the human body to keep normal levels of fluid outside cells

(Emsley, 2001; Stwertka, 2002)

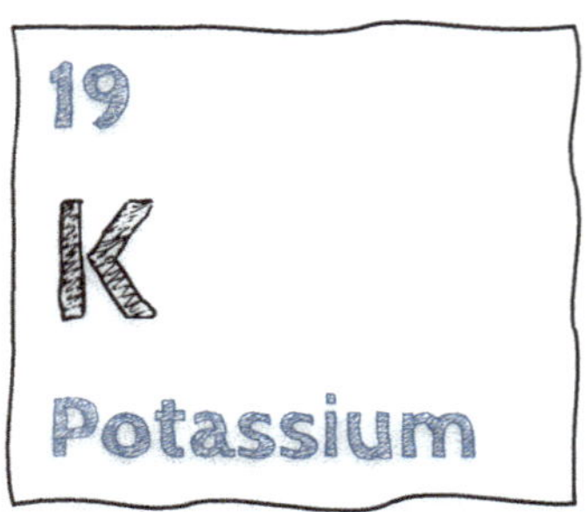

What item around you contains Potassium?

Draw it!

Write a short story that features the
Potassium element.

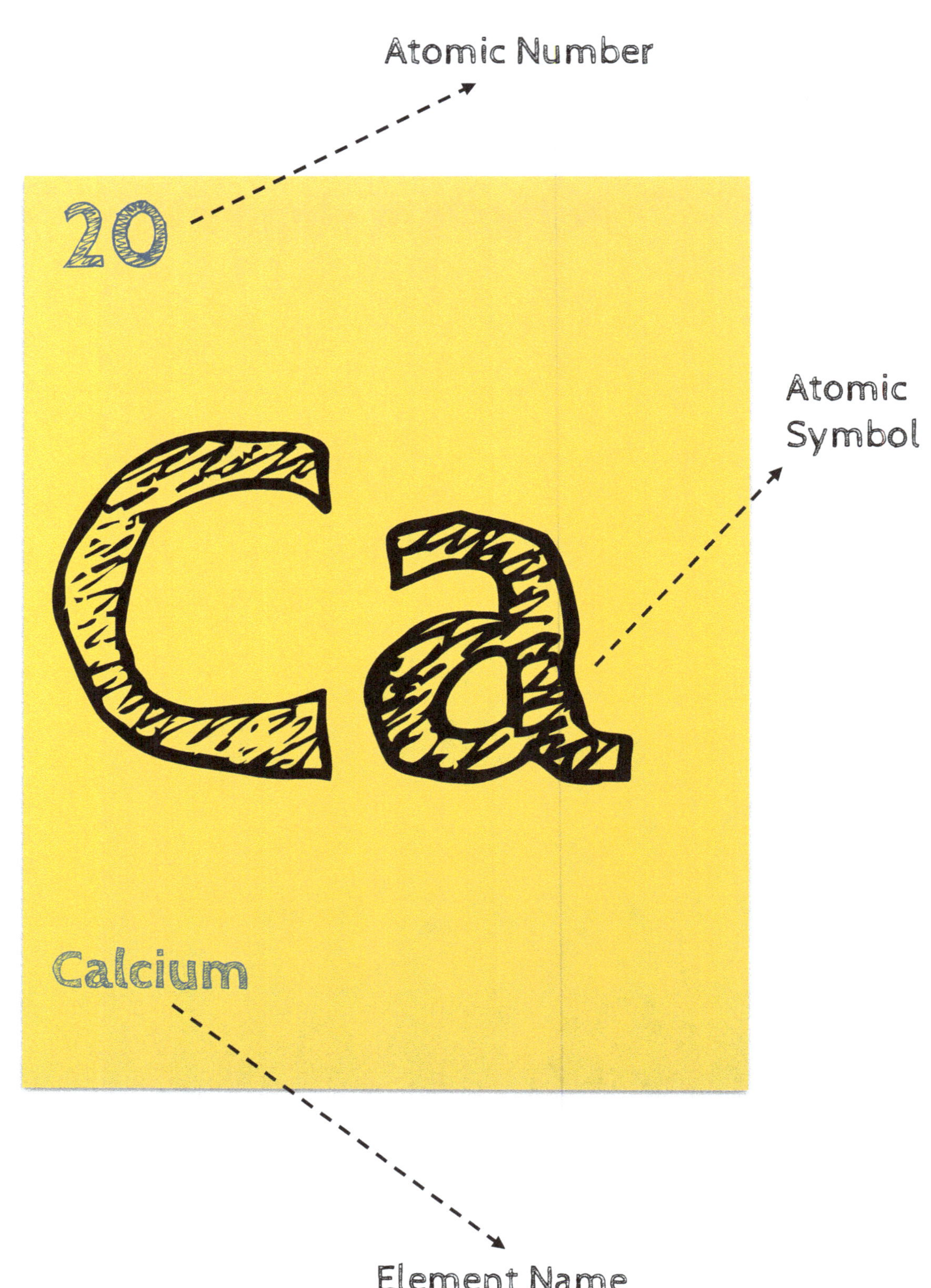

Atomic Number
20
Ca
Calcium
Atomic Symbol
Element Name

FACTS ABOUT CALCIUM

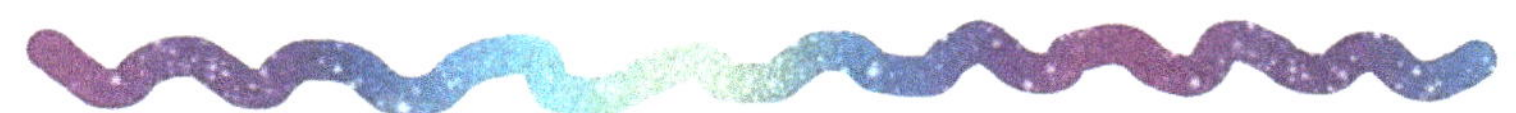

* Calcium is a soft metal that is silvery-white in color

* In the human body, calcium is found in bones and teeth

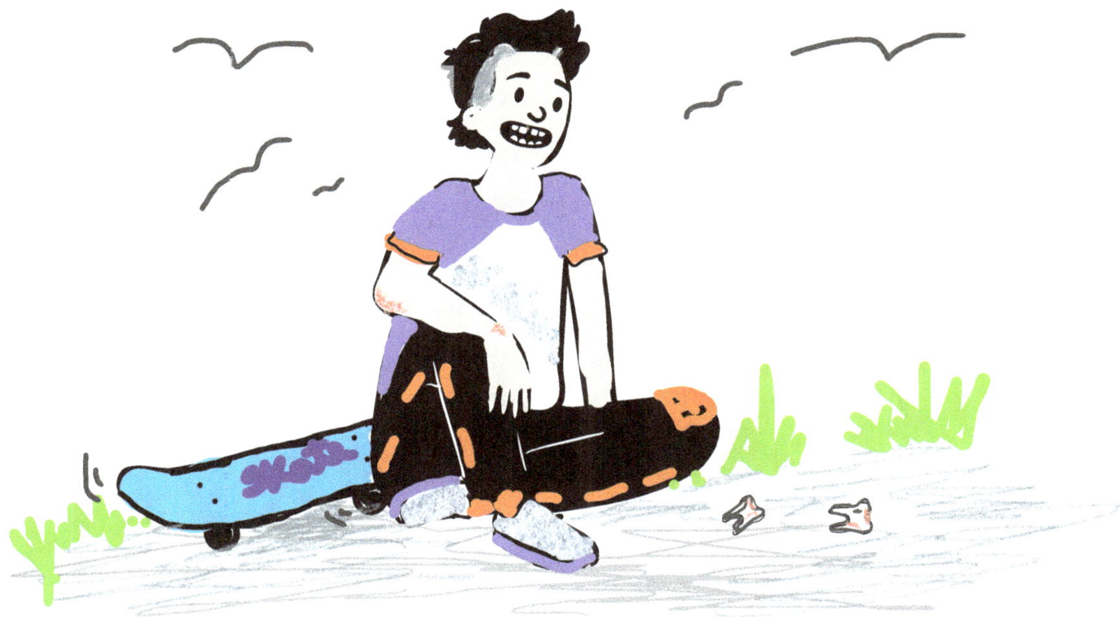

* Calcium is used to make cement for construction projects, such as schools, hospitals, and homes

(Emsley, 2001; Stwertka, 2002)

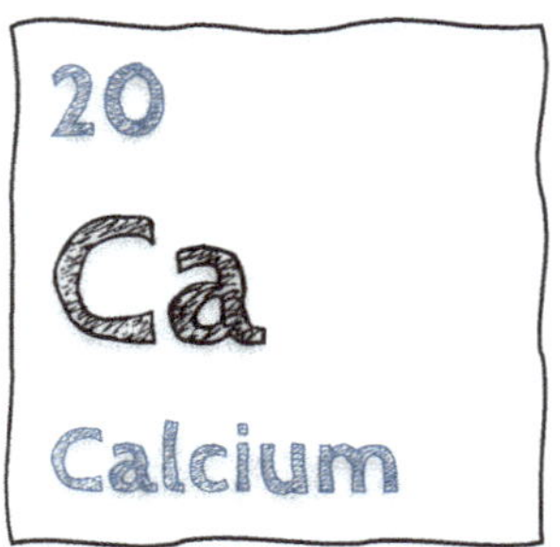

What item around you contains Calcium?

Draw it!

Write a short story that features the
Calcium element.

REFERENCES

Emsley, J. (2011). *Nature's building blocks: An AZ guide to the elements*. Oxford University Press.

Periodic Table of Elements - American Chemical Society. (n.d.). American Chemical Society. https://www.acs.org/education/whatische mistry/periodictable.html

Stwertka, A. (2002). *A Guide to the Elements*. Oxford University Press.